Metallurgy and Materials Science

a series of student texts

General Editors:

Professor R W K Honeycombe
Professor P Hancock

Steels

Microstructure and Properties

R W K Honeycombe

Goldsmiths' Professor of Metallurgy
University of Cambridge

Edward Arnold

© R W K Honeycombe 1981

First published in 1981
by Edward Arnold (Publishers) Ltd
41 Bedford Square, London WC1B 3DQ

British Library Cataloguing in Publication Data
Honeycombe, Robert William Kerr
   Steels. – (Metallurgy and materials science).
   1. Steel
   I. Title II. Series
   669'. 142   TN730

ISBN 0 7131 2793 7

To June

All Rights Reserved. No part of this publication may be reproduced, stored in a retrieval system, or transmitted in any form or by any means, electronic, mechanical, photocopying, recording or otherwise, without the prior permission of Edward Arnold (Publishers) Ltd.

Typeset by The Macmillan Company of India Ltd., Bangalore 1.
Printed and bound by Spottiswoode Ballantyne Ltd., Colchester.

General editors' preface

Large textbooks with broad subject coverage still have their place in university teaching. However, staff and students alike are attracted to compact, cheaper books which cover well-defined parts of a subject up to and beyond final year undergraduate work. The aim of this series is to do just this in metallurgy and materials science.

The subject as taught is gradually becoming more integrated and less polarized towards metals or non-metallic materials, so we have been able to plan a group of texts concerned with basic aspects of the subject, e.g. thermodynamics, phase transformations, deformation and fracture, which will comprise one part of the series. The other books, which are concerned specifically with the main groups of materials and processing routes of engineering interest, will use the basic principles to examine and explain materials behaviour over a wide range of conditions. The aim is not to cover each subject in the greatest depth, but to provide the student with a compact treatment as a springboard to further detailed studies, when he has chosen his particular field of work after graduation. Adequate general references will be provided for further study. The books are aimed not only at metallurgists and materials scientists, but also towards engineers and scientists wishing to know more about the structure and properties of engineering materials.

1980   RWKH
                                                                       PH

Preface

In this book, I have attempted to outline the principles which determine the microstructures of steels and through these the mechanical properties. At a time when our metallographic techniques are reaching almost to atomic resolution, it is essential to emphasize structure on the finest scale, especially because mechanical properties are sensitive to changes at this level. While this is not a book on the selection of steels for different uses, I have tried to include sufficient information to describe how broad categories of steels fulfill practical requirements. However, the main thrust of the book is to examine analytically how the γ/α phase transformation is utilized, and to explain the many effects that non-metallic and metallic alloying elements have, both on this transformation and on other phenomena.

This book is written with the needs of metallurgists, materials scientists and engineers in mind, and should be useful not only in the later years of the first degree and diploma courses but also in postgraduate courses. An elementary knowledge of materials science, metallography, crystallography and physics is assumed.

I am indebted to several colleagues for their interest in this book, particularly Dr D V Edmonds, who kindly read the manuscript, Dr P R Howell, Dr B Muddle and Dr H Bhadeshia, who made helpful comments on various sections, and numerous other numbers of my research group who have provided illustrations. I wish also to thank my colleagues in different countries for their kind permission to use diagrams from their work. I am also very grateful to Mr S D Charter for his careful preparation of the line diagrams. Finally, my warmest thanks go to Mrs Diana Walker and Miss Rosemary Leach for their careful and dedicated typing of the manuscript.

Cambridge RWKH
1980

Contents

	General editors' preface	v
	Preface	vii
1	**Iron and its interstitial solid solutions**	1
1.1	Introduction	1
1.2	The phase transformation: α- and γ-iron	1
1.3	Carbon and nitrogen in solution in α- and γ-iron	4
1.4	Some practical aspects	10
	Further reading	11
2	**The strengthening of iron and its alloys**	12
2.1	Introduction	12
2.2	Work hardening	12
2.3	Solid solution strengthening by interstitials	16
2.4	Substitutional solid solution strengthening of iron	21
2.5	Grain size	22
2.6	Dispersion strengthening	24
2.7	An overall view	25
2.8	Some practical aspects	25
	Further reading	27
3	**The iron-carbon equilibrium diagram and plain carbon steels**	28
3.1	The iron-carbon equilibrium diagram	28
3.2	The austenite-ferrite transformation	31
3.3	The austenite-cementite transformation	33
3.4	The kinetics of the γ/α transformation	34
3.5	The austenite-pearlite reaction	38
3.6	Ferrite-pearlite steels	50
	Further reading	53
4	**The effects of alloying elements in iron-carbon alloys**	55
4.1	The γ- and α-phase fields	55
4.2	The distribution of alloying elements in steels	58
4.3	The effect of alloying elements on the kinetics of the γ/α transformation	62

x *Contents*

4.4	Structural changes resulting from alloying additions	67
4.5	Transformation diagrams for alloy steels	74
	Further reading	75

5 The formation of martensite — 76
5.1	Introduction	76
5.2	General characteristics	76
5.3	The crystal structure of martensite	79
5.4	The crystallography of martensitic transformations	81
5.5	The morphology of ferrous martensites	86
5.6	Kinetics of transformation of martensite	90
5.7	The strength of martensite	98
	Further reading	105

6 The bainite reaction — 106
6.1	Introduction	106
6.2	Morphology and crystallography of upper bainite	106
6.3	Morphology and crystallography of lower bainite	108
6.4	Reaction kinetics of bainite formation	111
6.5	Role of alloying elements	114
6.6	Use of bainitic steels	117
	Further reading	119

7 The heat treatment of steels–hardenability — 121
7.1	Introduction	121
7.2	Use of TTT and continuous cooling diagrams	121
7.3	Hardenability testing	125
7.4	Effect of grain size and chemical composition on hardenability	130
7.5	Jominy tests and continuous cooling diagrams	133
7.6	Hardenability and heat treatment	135
7.7	Quenching stresses and quench cracking	137
7.8	Martempering	139
	Further reading	139

8 The tempering of martensite — 140
8.1	Introduction	140
8.2	Tempering of plain carbon steels	140
8.3	Mechanical properties of tempered plain carbon steels	147
8.4	Tempering of alloy steels	148
8.5	Maraging steels	164
	Further reading	165

9 Thermomechanical treatment of steels — 166
| 9.1 | Introduction | 166 |
| 9.2 | Controlled rolling of low alloy steels | 166 |

9.3	Dual phase steels	176
9.4	Ausforming	177
9.5	Isoforming	182
9.6	High temperature thermomechanical treatments (HTMT)	182
9.7	Industrial steels subjected to thermomechanical treatments	183
	Further reading	185
10	**The embrittlement and fracture of steels**	186
10.1	Introduction	186
10.2	Cleavage fracture in iron and steel	186
10.3	Factors influencing the onset of cleavage fracture	188
10.4	Criterion for the ductile/brittle transition	191
10.5	Practical aspects of brittle fracture	194
10.6	Ductile or fibrous fracture	196
10.7	Intergranular embrittlement	202
	Further reading	210
11	**Austenitic steels**	211
11.1	Introduction	211
11.2	The iron-chromium-nickel system	211
11.3	Chromium carbide in Cr-Ni austenitic steels	215
11.4	Precipitation of niobium and titanium carbides	219
11.5	Nitrides in austenitic steels	222
11.6	Intermetallic precipitation in austenite	223
11.7	Austenitic steels in practical applications	224
11.8	Duplex and ferritic stainless steels	227
11.9	The transformation of metastable austenite	230
	Further reading	234
	Index	237

1
Iron and its interstitial solid solutions

1.1 Introduction

The study of steels is important because steels represent by far the most widely-used metallic materials, primarily due to the fact that they can be manufactured relatively cheaply in large quantities to very precise specifications. They also provide an extensive range of mechanical properties from moderate strength levels (200–300 MN m^{-2}) with excellent ductility and toughness, to very high strengths (2000 MN m^{-2}) with adequate ductility. It is, therefore, not surprising that irons and steels comprise well over 80% by weight of the alloys in general industrial use. This book aims to explain why steels take this pre-eminent position, and examines in detail the phenomena whose exploitation enables the desired properties to be achieved.

Steels form perhaps the most complex group of alloys in common use. Therefore, in studying them it is useful to consider the behaviour of pure iron first, then iron-carbon alloys, and finally examine the many complexities which arise when further alloying additions are made. Pure iron is not an easy material to produce. However, it has recently been made with a total impurity content not exceeding 60 ppm (parts per million), of which 10 ppm is accounted for by non-metallic impurities such as carbon, oxygen, sulphur, phosphorus, while 50 ppm represents the metallic impurities. Iron of this purity is extremely weak: the resolved shear stress of a single crystal at room temperature can be as low as 10 MN m^{-2}, while the yield stress of a polycrystalline sample at the same temperature can be well below 150 MN m^{-2}. Some of the mechanisms which raise iron to the strength levels associated with steels will be discussed, before the wide range of complex structures which determine the properties of steels is dealt with.

1.2 The phase transformation: α- and γ- iron

Pure iron exists in two crystal forms, one body-centred cubic (bcc) (α-iron, ferrite) which remains stable from low temperatures up to 910 °C (the A_3 point), when it transforms to a face-centred cubic (fcc) form (γ-iron, austenite). The γ-iron remains stable until 1390 °C, the A_4 point, when it reverts to bcc form, (now δ-iron) which remains stable up to the melting point of 1536°C. Fig. 1.1 shows the phase changes in a plot of the mean

2 Steels—Microstructure and Properties

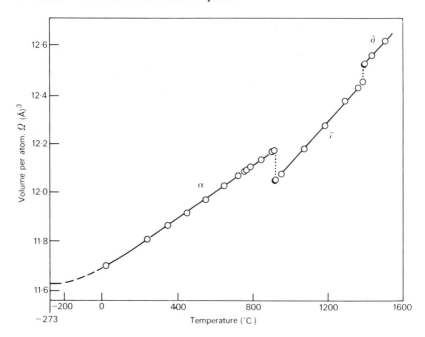

Fig. 1.1 Temperature dependence of the mean volume per atom in iron crystals (Hume-Rothery, *The Structure of Alloys of Iron*, Pergamon, 1966)

volume per atom of iron as a function of temperature. It should be noted that the γ to α transformation is accompanied by an atomic volume change of approximately 1%, which leads to the generation of internal stresses during transformation.

The detailed geometry of unit cells of α- and γ-iron crystals is particularly relevant to, for example, the solubility in the two phases of non-metallic elements such as carbon and nitrogen, the diffusivity of alloying elements at elevated temperatures, and the general behaviour on plastic deformation. The bcc structure of α-iron is more loosely packed than that of fcc γ-iron (Figs. 1.2a, b). The largest cavities in the bcc structure are the tetrahedral holes existing between two edge and two central atoms in the structure, which together form a tetrahedron (Fig. 1.2c). The second largest are the octahedral holes which occupy the centres of the faces and the $\langle 001 \rangle$ edges of the body-centred cube (Fig. 1.2d). The surrounding iron atoms are at the corners of a flattened octahedron (Fig. 1.2e). It is interesting that the fcc structure, although more closely-packed, has larger holes than the bcc structure. These holes are at the centres of the cube edges, and are surrounded by six atoms in the form of an octagon, so they are referred to as octahedral holes (Fig. 1.2f). There are also smaller tetrahedral interstices. The largest diameters of spheres which will enter these interstices are given in Table 1.1.

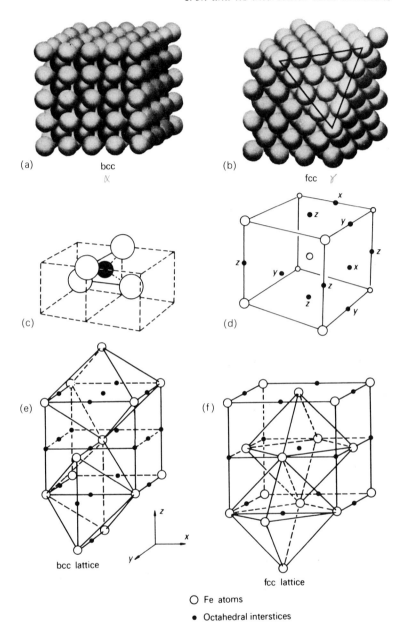

Fig. 1.2 a, Body-centered cubic structure, b, face-centred cubic structure (Moffat, Pearsall and Wulff, *The Structure and Properties of Materials: Vol. 1, Structure*, John Wiley); c, tetrahedral interstices in bcc structure, d, octahedral interstices in bcc structure (Hume-Rothery, *The Structure of Alloys of Iron*, Pergamon, 1966); e and f, octahedral interstices in bcc and fcc iron (Cohen, *Trans. Met. Soc. AIME*, 1962, **224**, 638)

4 *Steels—Microstructure and Properties*

Table 1.1 Size of largest spheres fitting interstices in bcc and fcc structures

		Diameter	Diameter in iron (Å)
bcc	tetrahedral	$0.29r$	0.36
	octahedral	$0.15r$	0.19
fcc	tetrahedral	$0.23r$	0.28
	octahedral	$0.41r$	0.51

r = atomic radius

The $\alpha \rightleftharpoons \gamma$ transformation in pure iron occurs very rapidly, so it is impossible to retain the high-temperature fcc form at room temperature. Rapid quenching can substantially alter the morphology of the resulting α-iron, but it still retains its bcc structure. It follows that any detailed study of austenite (γ-iron) in pure iron must be done dynamically at elevated temperatures, e.g. by high temperature X-ray diffraction, while the transformation to ferrite (α-iron) can be best detected by precision dilatometry, which shows the phase transformation as a result of the volume change.

1.3 Carbon and nitrogen in solution in α- and γ-iron

1.3.1 Solubility of carbon and nitrogen in α- and γ-iron

The addition of carbon to iron is sufficient to form a steel. However, steel is a generic term which covers a very large range of complex compositions. The presence of even a small concentration of carbon, e.g. 0.1–0.2 weight per cent (wt%); approximately 0.5–1.0 atomic per cent (at%), has a great strengthening effect on iron, a fact known to smiths over 2 500 years ago since iron heated in a charcoal fire can readily absorb carbon by solid state diffusion. However, the detailed processes by which the absorption of carbon into iron converts a relatively soft metal into a very strong and often tough alloy have only recently been fully explored.

The atomic sizes of carbon and nitrogen (Table 1.2) are sufficiently small relative to that of iron to allow these elements to enter the α-iron and γ-iron lattices as *interstitial* solute atoms. In contrast, the metallic alloying elements such as manganese, nickel and chromium have much larger atoms, i.e. nearer in size to those of iron, and consequently they enter into *substitutional* solid solution. However, comparison of the atomic sizes of C and N with the sizes of the available interstices makes it clear that some lattice distortion must take place when these atoms enter the iron lattice. Indeed, it is found that C and N in α-iron occupy not the larger tetrahedral holes, but the octahedral interstices which are more favourably placed for the relief of strain, which occurs by movement of two nearest neighbour

Table 1.2 Atomic sizes of non-metallic elements in iron

Element	Atomic radius, r (Å)	$\dfrac{r}{r_{Fe}}$
α-Fe	1.28	1.00
B	0.94	0.73
C	0.77	0.60
N	0.72	0.57
O	0.60	0.47
H	0.46	0.36

iron atoms. In the case of tetrahedral interstices, four iron atoms are of nearest-neighbour status and the displacement of these would require more strain energy. Consequently these interstices are not preferred sites for carbon and nitrogen atoms.

The solubility of both C and N in austenite should be greater than in ferrite, because of the larger interstices available. Table 1.3 shows that this is so for both elements, the solubility in γ-iron rising as high as 9–10 at %, in contrast to the maximum solubility of C in α-iron of 0.1 at % and of N in α-iron of 0.4 at %. These marked differences of the solubilities of the main interstitial solutes in γ- and in α-iron are of profound significance in the heat treatment of steels, and are fully exploited to increase strength (see Chapter 2). It should be noted that the room temperature solubilities of both C and N in α-iron are extremely low, well below the actual interstitial contents of many pure irons. It is, therefore, reasonable to expect that during simple heat treatments, excess carbon and nitrogen will be precipitated. This could happen in heat treatments involving quenching from the γ state, or even after treatments entirely within the α field, where the solubility of C varies by nearly three orders of magnitude between 720°C and 20°C.

Table 1.3 Solubilities of carbon and nitrogen in γ- and α-iron

	Temperature (°C)	Solubility wt%	at%
C in γ-iron	1150	2.04	8.8
	723	0.80	3.6
C in α-iron	723	0.02	0.095
	20	<0.00005	<0.00012
N in γ-iron	650	2.8	10.3
	590	2.35	8.75
N in α-iron	590	0.10	0.40
	20	<0.0001	<0.0004

Fortunately, sensitive physical techniques allow the study of small concentrations of interstitial solute atoms in α-iron. Snoek first showed that internal friction measurements on iron wire oscillating in a torsional pendulum, over a range of temperature just above ambient temperature, revealed an energy loss peak (Snoek peak) at a particular temperature for a given frequency. It was shown that the energy loss was associated with the migration of carbon atoms from randomly chosen octahedral interstices to those holes which were enlarged on application of the stress in one direction, followed by a reverse migration when the stress changed direction and made other interstices larger. This movement of carbon atoms at a critical temperature is an additional form of damping or internal friction: below the critical temperature the diffusivity is too small for atomic migration, and above it the migration is too rapid to lead to appreciable damping. The height of the Snoek peak is proportional to the concentration of interstitial atoms, so the technique can be used not only to determine the very low solubilities of interstitial elements in iron, but also to examine the precipitation of excess carbon or nitrogen during an ageing treatment.

1.3.2 Diffusion of solutes in iron

The internal friction technique can also be used to determine the diffusivities of C and N in α-iron (Table 1.4). The temperature dependence of diffusivity follows the standard exponential relationship:

$$D_C = 0.02 \exp\left(-\frac{Q}{RT}\right) \text{cm}^2 \text{ s}^{-1} \qquad (Q = 80 \text{ kJ mol}^{-1})$$

$$D_N = 6.6 \times 10^{-3} \exp\left(-\frac{Q}{RT}\right) \text{cm}^2 \text{ s}^{-1} \qquad (Q = 76 \text{ kJ mol}^{-1})$$

where D_C and D_N are the diffusion coefficients of carbon and nitrogen, respectively, and Q is the activation energy. The dependence of D_C and D_N on temperature is shown graphically in Fig. 1.3.

Different techniques, e.g. involving radioactive tracers, have to be used for substitutional elements. A comparison of the diffusivities of the interstitial atoms with those of substitutional atoms, i.e. typical metallic solutes, on both α- and γ-iron, shows that the substitutional atoms move several orders of magnitude more slowly (Table 1.4). This is a very important distinction which is relevant to some of the more complex phenomena in alloy steels. However, for the time being, it should be noted that homogenizing treatments designed to eliminate concentration gradients of solute elements need to be much more prolonged and at higher temperatures when substitutional rather than interstitial solutes are involved.

The other major point which is illustrated in Table 1.4 is that, for a

Table 1.4 Diffusivities of elements in γ- and α-iron

Solvent	Solute	Activation energy, Q (kJ mol^{-1})	Frequency factor, D_0 (cm^2 s^{-1})	Diffusion coefficient, $D_{910°C}$ (cm^2 s^{-1})	Temperature range (°C)
γ-iron	C	135	0.15	1.5×10^{-7}	900–1050
	Fe	269	0.18	2.2×10^{-13}	1060–1390
	Co	364	3.0×10^2	24.0×10^{-12} (at 1050°C)	1050–1250
	Cr	405	1.8×10^4	58.0×10^{-12} (at 1050°C)	1050–1250
	Cu	253	3.0	15.0×10^{-11}	800–1200
	Ni	280	0.77	7.7×10^{-13}	930–1050
	P	293	28.3	3.6×10^{-12}	1280–1350
	S	202	1.35	1.5×10^{-9}	1200–1350
	W	376	1.0×10^3	12.0×10^{-12} (at 1050°C)	1050–1250
α-iron	C	80	6.2×10^{-3}	1.8×10^{-6}	
	N	76	3.0×10^{-3}	1.3×10^{-6}	
	Fe	240	0.5		700–750
	Co	226	0.2	2.1×10^{-11}	700–790
	Cr	343	3.0×10^4		
	Ni	258	9.7	3.7×10^{-11}	700–900
	P	230	2.9	2.0×10^{-10}	860–900
	W	293	3.8×10^2		

Data from Askill, J., *Tracer Diffusion Data for Metals, Alloys and Simple Oxides*, IFI/Plenum, 1970; Wohlbier, F. H., *Diffusion and Defect Data*, Materials Review Series, Vol. 12, Nos. 1–4, Trans Tech Publications, 1976; Krishtal, M. A., *Diffusion Processes in Iron Alloys*, translated from Russian by A. Wald, ed. J. J. Becker, Israel Program for Scientific Translations, Jerusalem, 1970.

particular temperature, diffusion of both substitutional and interstitial solute occurs much more rapidly in ferrite than in austenite. This arises because γ-iron is a close-packed structure whereas α-iron, which is more loosely-packed, responds more readily to thermal activation and allows easier passage through the structure of vacancies and associated solute atoms. In all cases, the activation energy Q is less for a particular element diffusing in α-iron, than it is for the same element diffusing in γ-iron.

1.3.3 Precipitation of carbon and nitrogen from α-iron

α-iron containing about 0.02 wt% C is substantially supersaturated with carbon if, after being held at 700°C, it is quenched to room temperature. This supersaturated solid solution is not stable, even at room temperature, because of the ease with which carbon can diffuse in α-iron. Consequently, in the range 20–300°C, carbon is precipitated as iron carbide. This process

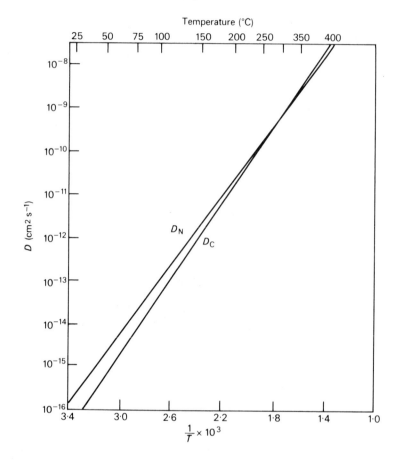

Fig. 1.3 Temperature dependence of diffusion coefficients of nitrogen (D_N) and carbon (D_C) in α-iron (Baird, *Iron and Steel*, Illiffe Production Publications, 1963)

has been followed by measurement of changes in physical properties such as electrical resistivity, internal friction, and by direct observation of the structural changes in the electron microscope.

The process of ageing is a two-stage one. The first stage takes place at temperatures up to 200°C and involves the formation of a transitional iron carbide phase (ε) with a close-packed hexagonal structure which is often difficult to identify, although its morphology and crystallography have been established. It forms as platelets on $\{100\}_\alpha$ planes, apparently homogenously in the α-iron matrix, but at higher ageing temperatures (150–200°C) nucleation occurs preferentially on dislocations. The composition is between $Fe_{2.4}C$ and Fe_3C. Ageing at 200°C and above leads to the second stage of ageing in which orthorhombic cementite Fe_3C is formed as platelets on $\{110\}_\alpha$ planes in $\langle 111 \rangle_\alpha$ directions. Often the platelets grow on several $\{110\}$ planes from a common centre giving rise to structures which appear dendritic in character (Fig. 1.4). The transition from ε-iron carbide to cementite is difficult to study, but it appears to occur by nucleation of cementite at the ε-carbide/α interfaces, followed by re-solution of the metastable ε-carbide precipitate.

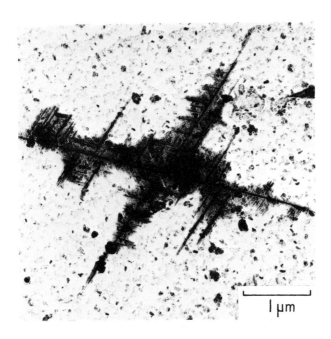

Fig. 1.4 Cementite precipitation in quench-aged iron, 1560 min at 240°C (Langer). Replica EM, ×18 000

The maximum solubility of nitrogen in ferrite is 0.10 wt %, so a greater volume fraction of nitride precipitate can be obtained. The process is again two-stage with a bc tetragonal α'' phase, $Fe_{16}N_2$, as the intermediate precipitate, forming as discs on $\{100\}_\alpha$ matrix planes both homogeneously and on dislocations. Above about 200°C, this transitional nitride is replaced by the ordered fcc γ', Fe_4N, which forms as platelets with $\{112\}_{\gamma'}//\{210\}_\alpha$

The ageing of α-iron quenched from a high temperature in the α-range is usually referred to as *quench ageing*, and there is substantial evidence to show that the process can cause considerable strengthening, even in relatively pure iron. Fig. 1.5 plots the hardness changes in an Fe-0.02 wt % nitrogen alloy, aged at 60°C after quenching from 500°C, which were shown by micro-examination to be due to precipitation of $Fe_{16}N_2$. In commercial low carbon steels, nitrogen is usually combined with aluminium, or present

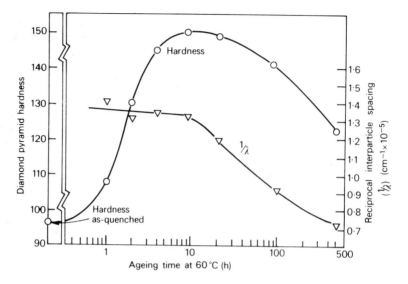

Fig. 1.5 Quench ageing of iron with 0.02 wt % N. Variations of hardness and particle spacing (λ) with ageing time at 60°C (Keh and Leslie, *Materials Science Research*, 1963, **1**)

in too low a concentration to make a substantial contribution to quench ageing, with the result that the major effect is due to carbon. This behaviour should be compared with that of *strain ageing* (see Section 2.3.1).

1.4 Some practical aspects

The very rapid diffusivity of carbon and nitrogen in iron compared with that of the metallic alloying elements is exploited in the processes of *carburizing* and *nitriding*. Carburizing can be carried out by heating a low carbon steel in contact with carbon to the austenitic range, e.g. 1000°C, where the carbon solubility, c_1, is substantial. The result is a carbon gradient in the steel, from c_1 at the surface in contact with the carbon, to c at a depth x. The solution of Fick's second diffusion law for this case is:

$$c = c_1 \left[1 - \text{erf}\left(\frac{x}{2\sqrt{Dt}}\right) \right] \tag{1.1}$$

which is essentially the equation of the concentration-depth curve, where t = time in seconds. The diffusion coefficient D of carbon in iron actually varies with carbon content, so the above relationship is not rigorously obeyed. Carburizing, whether carried out using carbon, or more efficiently using a carburizing gas (*gas carburizing*), provides a high carbon surface on a steel, which, after appropriate heat treatment, is strong and wear resistant.

Nitriding is normally carried out in an atmosphere of ammonia, but at a lower temperature (500–550°C) than carburizing, consequently the reaction occurs in the ferrite phase, in which nitrogen has a substantially higher solubility than carbon. Nitriding steels usually contain chromium ($\sim 1\%$), aluminium ($\sim 1\%$), vanadium or molybdenum ($\sim 0.2\%$), which are nitride-forming elements, and which contribute to the very great hardness of the surface layer produced.

In cast steels, metallic alloying elements are usually segregated on a microscopic scale, by *coring of dendrites*. Therefore, to obtain a more uniform distribution, *homogenization* annealing must be carried out, otherwise the inhomogeneities will persist even after large amounts of mechanical working. The much lower diffusivities of the metallic alloying elements compared with carbon and nitrogen, means that the homogenization must be carried out at very high temperatures (1200–1300°C), approaching the melting point, hence the use of *soaking pits* where steel ingots are held after casting and prior to hot rolling. The higher the alloying element content of the steel, the more prolonged must be this high temperature treatment.

Further reading

Hume-Rothery, W., *The Structures of Alloys of Iron*, Pergamon Press, 1966

Goldschmidt, H. J., *Interstitial Alloys*, Butterworths, 1967

American Society for Metals, *Thermodynamics in Physical Metallurgy* (particularly chapter by C. A. Wert), 1950

Speich, G. R. and Clark, J. B. (eds), *Precipitation from Iron-Base Alloys*, Met. Soc. AIME Conference, Gordon and Breach, 1965

Krishtal, M. A., *Diffusion Processes in Iron Alloys*, translated from Russian by A. Wald, ed. J. J. Becker, Israel Program for Scientific Translations, Jerusalem, 1970

2
The strengthening of iron and its alloys

2.1 Introduction

Although pure iron is a weak material, steels cover a wide range of the strength spectrum from low yield stress levels (around 200 MN m^{-2}) to very high levels (approaching 2000 MN m^{-2}). These mechanical properties are usually achieved by the combined use of several strengthening mechanisms, and in such circumstances it is often difficult to quantify the different contributions to the strength. Consequently, the several basic ways in which iron can be strengthened will be discussed, illustrating the phenomena by reference to simple iron-base systems. These results should then be helpful in examining the behaviour of more complex steels.

Like other metals, iron can be strengthened by several basic mechanisms, the most important of which are:
(1) Work hardening
(2) Solid solution strengthening by interstitial atoms
(3) Solid solution strengthening by substitutional atoms
(4) Refinement of grain size
(5) Dispersion strengthening, including lamellar and random dispersed structures.

The most distinctive aspect of strengthening of iron is the role of the interstitial solutes carbon and nitrogen. These elements also play a vital part in interacting with dislocations, and in combining preferentially with some of the metallic alloying elements used in steels.

2.2 Work hardening

Work hardening is an important strengthening process in steel, particularly in obtaining high strength levels in rod and wire, both in plain carbon and alloy steels. For example, the tensile strength of an 0.05% C steel subjected to 95% reduction in area by wire drawing, is raised by no less than 550 MN m^{-2}, while higher carbon steels are strengthened by up to twice this amount. Indeed, without the addition of special alloying elements, plain carbon steels can be raised to strength levels above 1500 MN m^{-2} simply by the phenomenon of work hardening.

Basic work on the deformation of iron has largely concentrated on the other end of the strength spectrum, namely pure single crystals and polycrystals subjected to small controlled deformations. This approach has shown that the slip plane in α-iron is not unique. Slip occurs on several planes, {110}, {112} and {123}, but always in the close packed ⟨111⟩ direction which is common to each of these planes (Fig. 2.1). The diversity of slip planes leads to rather irregular wavy slip bands in deformed crystals, as the dislocations can readily move from one type of plane to another by cross slip, provided they share a common slip direction. The Burgers vector of the slip dislocations would thus be expected to be $\frac{1}{2}\langle 111 \rangle$, which has been confirmed by thin-foil electron microscopy.

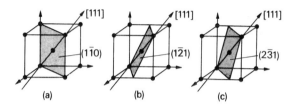

Fig. 2.1 The slip systems in the bcc structure

The yield stress of iron single crystals is very sensitive to both temperature and strain rate, and a similar dependence has been found for less pure polycrystalline iron. Fig. 2.2 shows the flow stress σ_T at temperature T, less that at room temperaure σ_{293}, plotted against T, showing that both single crystal and polycrystalline iron of different interstitial content give values falling on the one curve. Therefore, the temperature sensitivity

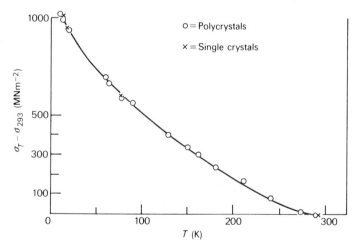

Fig. 2.2 Temperature dependence of the flow stress of single crystals and polycrystals of pure iron (Christian, *Phil. Trans. Roy. Soc.*, 1967, **261**, 253)

cannot be attributed to interstitial impurities. It is explained by the effect of temperature on the stress needed to move free dislocations in the crystal, the Peierls-Nabarro stress. Direct observation of screw dislocations in iron in the electron microscope has shown that their ease of mobility decreases strongly with decreasing temperature.

If the shear stress at any point on the stress-strain curve is considered, the measured shear stress τ for further deformation comprises two quantities:

$$\tau = \tau^* + \tau_i \tag{2.1}$$

The effective shear stress, τ^* arises from the interaction of the dislocations with short range obstacles, e.g. isolated dislocations. This stress is strongly temperature-dependent as thermal activation is helpful in moving dislocations around short range obstacles†. On the other hand, τ_i is the internal stress arising from long range obstacles such as grain boundaries, cell walls and other complex dislocation arrays††. In these circumstances thermal fluctuations are of no assistance. The two component stresses are defined as follows:

$$\tau^* = \frac{1}{V}\left[\Delta H_0 + kT\ln\frac{l\dot{\varepsilon}}{\rho m A b \gamma}\right] \tag{2.2}$$

$$\tau_i = \alpha \mu b \rho^2 \tag{2.3}$$

where V = activation volume
ΔH_0 = activation enthalpy at $\tau^* = 0$
k = Boltzmann's constant
T = temperature
l = length of dislocation line activated
$\dot{\varepsilon}$ = strain rate
m = mobile dislocation density
A = area of glide plane covered by dislocation
b = Burgers vector
γ = frequency of vibration of dislocation line length
α = constant
μ = shear modulus
ρ = dislocation density

The initial flow stress or yield stress with its large temperature dependence arises primarily from τ^*, while the increment in flow stress resulting from work hardening is largely independent of temperature, and is caused by the increase in τ_i with increasing strain as the dislocation density ρ increases.

† Conrad, K., *J.I.S.I.*, 1961, **198**, 364
†† Michalak, J. J., *Acta Met.*, 1965, **13**, 213

2.3 Solid solution strengthening by interstitials

2.3.1 The yield point

Carbon and nitrogen, even in concentrations as low as 0.005 wt %, in iron leads to a sharp transition between elastic and plastic deformation in a tensile test (Fig. 2.3a). Decarburization of the iron results in the elimination of this sharp transition or *yield point*, which implies that the solute atoms are in some way responsible for this striking behaviour. Frequently the load drops dramatically at the *upper yield point* (A) to another value referred to as the *lower yield point* (B). Under some experimental conditions, yield drops of about 30 % of the upper yield stress can be obtained. Following the lower yield point, there is frequently a horizontal section of the stress-strain curve (BC) during which the plastic deformation propagates at a front which can move uniformly along the specimen. This front is referred to as a *Luders band* (Fig. 2.3b), and the horizontal portion, BC, of the stress-strain curve as the Luders extension. The development of Luders bands can be much less uniform, and, for example, in pressings where the stress is far from uniaxial, complex arrays of bands can be observed. These are often referred to as *stretcher strains*, but they are still basically Luders bands. When the whole specimen has yielded, general work hardening commences and the stress-strain curve begins to rise in the normal way. If, however, this deformation is interrupted, and the specimen allowed to rest either at room temperature or for a shorter time at $100-150°C$, on reloading a new yield point is observed (D). This return of the yield point is referred to as *strain ageing*.

2.3.2 The role of interstitial elements in yield phenomena

The sharp upper and lower yield point in iron is eliminated by annealing in wet hydrogen, which reduces the carbon and nitrogen to very low levels. However, substantial strain ageing can occur at carbon levels around 0.002 wt %, and as little as 0.001–0.002 wt % N can result in severe strain ageing. Nitrogen is more effective in this respect than carbon, because its residual solubility near room temperature is substantially greater than that of carbon (Table 1.3).

Cottrell and Bilby first showed that interstitial atoms such as carbon and nitrogen would interact strongly with the strain fields of dislocations. The interstitial atoms have strain fields around them, but when such atoms move within the dislocation strain fields, there should be an overall reduction in the total strain energy. This leads to the formation of interstitial concentrations or *atmospheres* in the vicinity of dislocations, which in an extreme case can amount to lines of interstitial atoms along the cores of the dislocations (*condensed atmospheres*), for example in edge dislocations at the region of the strain field where there is maximum dilation

16 *Steels—Microstructure and Properties*

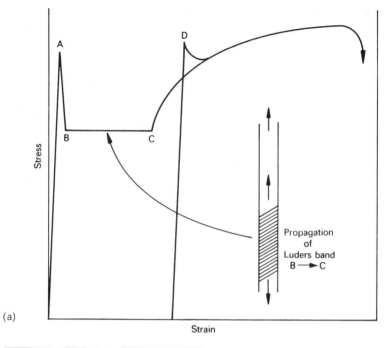

(a)

(b)

(c)

Fig. 2.3 a, Schematic diagram of yield phenomena as shown in a tensile test; b, Luders bands in deformed steel specimens (Hall, *Yield Point Phenomena in Metals and Alloys*, Macmillan, 1970); c, Luders bands in a notched steel specimen

(Figs. 2.4a and b). The binding energy between a dislocation in iron and a carbon atom is about 0.5 eV. Consequently dislocations can be locked in position by strings of carbon atoms along the dislocations, thus substantially raising the stress which would be necessary to cause dislocation movement. A particular attraction of this theory is that only a very small concentration of interstitial atoms is needed to produce locking along the whole length of all dislocation lines in annealed iron. For a typical dislocation density of 10^8 lines cm^{-2} in annealed iron, a carbon concentration of 10^{-6} wt% would be sufficient to provide one interstitial carbon atom per atomic plane along all the dislocation lines present, i.e. to saturate the dislocations. Consequently, this theory can explain the observation of yield phenomena at very low carbon and nitrogen concentrations.

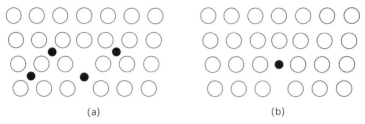

Fig. 2.4 Interstitial atoms in the vicinity of an edge dislocation: a, random atmosphere; b, condensed

The formation of interstitial atmospheres at dislocations requires diffusion of the solute. As both carbon and nitrogen diffuse very much more rapidly in iron than substitutional solutes, it is not surprising that strain ageing can take place readily in the range 20°C to 150°C. The interstitial concentration, c, in a dislocation strain field at a point where the binding energy is U is given by:

$$c = c_0 U/kT \qquad (2.4)$$

where c_0 is the average concentration. In general, this approach leads to a Maxwellian distribution of solute about the dislocation, but for carbon and nitrogen in steel, the elastic interaction energy U between solute and dislocation is so large that $U \gg kT$. Consequently the atmosphere condenses to form rows of interstitial atoms along the cores of the dislocations.

The critical temperature T_{crit} below which there is condensation of the atmosphere, occurs when $c = 1$ and $U = U_{max}$:

$$T_{crit} = \frac{U_{max}}{k \ln \dfrac{1}{c_0}} \qquad (2.5)$$

18 *Steels—Microstructure and Properties*

Therefore, if $c_0 = 10^{-4}$ and $U_{max} \simeq 10^{-19} J$, $T_{crit} = 700$ K. Thus the yield point would be expected at temperatures below 700 K, but it should disappear at higher temperatures when dislocations can escape from their atmospheres as a result of thermal activation. This corresponds approximately with results of experiments showing the temperature dependence of the stress-strain curve of mild steel (Fig. 2.5). As expected, the zone of yielding is well defined at the lower testing temperatures, becoming less regular as the temperature is raised, until it is replaced by fine serrations along the whole stress-strain curve. This phenomenon is referred to as *dynamic strain ageing*, in which the serrations represent the replacement of the primary yield point by numerous localized yield points within the specimen. These arise because the temperature is high enough to allow interstitial atoms to diffuse during deformation, and to form atmospheres around dislocations generated throughout the stress-strain curve. Steels tested under these conditions also show low ductilities, due partly to the high dislocation density and partly to the nucleation of carbide particles on

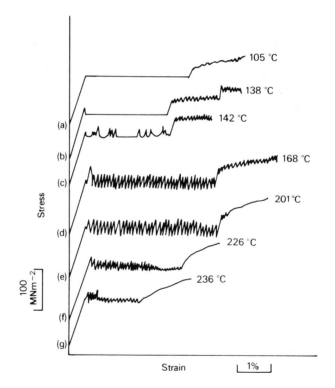

Fig. 2.5 Typical stress-strain curves for mild steel at elevated temperatures (Hall, *Yield Point Phenomena in Metals and Alloys*, Macmillan, 1970)

the dislocations where the carbon concentration is high. The phenomenon is often referred to as *blue brittleness*, blue being the interference colour of the steel surface when oxidized in this temperature range.

The break away of dislocations from their carbon atmospheres as a cause of the sharp yield point became a controversial aspect of the theory because it was found that the provision of free dislocations, for example, by scratching the surface of a specimen, did not eliminate the sharp yield point. An alternative theory was developed which assumed that, once condensed carbon atmospheres are formed in iron, the dislocations remain locked, and the yield phenomena arise from the generation and movement of newly-formed dislocations (Gilman and Johnson). The velocity of movement v of these fresh dislocations is related to the applied stress as follows:

$$v = \left(\frac{\sigma}{\sigma_r}\right)^m \tag{2.6}$$

where σ_r is a reference stress, σ is the yield stress, and m is an index characteristic of the material, varying between 1 and 60. The strain rate $\dot{\varepsilon}$ can be defined in terms of the movement of dislocations, as

$$\dot{\varepsilon} = nvb \tag{2.7}$$

where n is the number of mobile dislocations per unit area, v is their average velocity and b is Burgers vector.

Using Equation (2.7) the strain rates at the upper yield point ($\dot{\varepsilon}_U$) and the lower yield point ($\dot{\varepsilon}_L$) can be defined as follows:

$$\dot{\varepsilon}_U = \rho_U v_U b$$

where ρ_U is the mobile dislocation density at upper yield point and

$$\dot{\varepsilon}_L = \rho_L v_L b$$

where ρ_L is the mobile dislocation density at lower yield point, so

$$\frac{v_U}{v_L} = \frac{\rho_L}{\rho_U}$$

and using Equation (2.6)

$$\frac{\sigma_U}{\sigma_L} = \left(\frac{\rho_L}{\rho_U}\right)^{\frac{1}{m}} \tag{2.8}$$

Thus the ratio $\frac{\sigma_U}{\sigma_L}$ will be large, i.e. there will be a large yield drop, when m is small, and when ρ_L is much larger than ρ_U. Consequently, if at the upper yield stress the density of mobile dislocations is low, for example as a result of solute atom locking, a large drop in yield stress will occur if a large number of new dislocations is generated. Observations indicate that the dislocation density just after the lower yield stress is much higher than that observed at the upper yield point.

To summarize, the occurrence of a sharp yield point depends on the occurrence of a sudden increase in the number of mobile dislocations. However, the precise mechanism by which this takes place will depend on the effectiveness of the locking of the pre-existing dislocations. If the pinning is weak, then the yield point can arise as a result of unpinning. However, if the dislocations are strongly locked, either by interstitial atmospheres or precipitates, the yield point will result from the rapid generation of new dislocations.

Under conditions of dynamic strain ageing, where atmospheres of carbon atoms form continuously on newly-generated dislocations, it would be expected that a higher density of dislocations would be needed to complete the deformation, if it is assumed that most dislocations which attract carbon atmospheres are permanently locked in position. Electron microscopic observations have shown that in steels deformed at 200°C, the dislocation densities are an order of magnitude greater than those in specimens similarly deformed at room temperature. This also accounts for the fact that increased work hardening rates are obtained in the blue-brittle condition as compared to tests at room temperature.

2.3.3 Strengthening at high interstitial concentrations

Austenite can take into solid solution up to 10 at % carbon which can be retained in solid solution by rapid quenching. However, in these circumstances the phase transformation takes place, not to ferrite but to a tetragonal structure referred to as *martensite* (see Chapter 5). This phase forms as a result of a diffusionless shear transformation leading to characteristic laths or plates, which normally appear acicular in polished and etched sections. If the quench is sufficiently rapid, the martensite is essentially a supersaturated solid solution of carbon in a tetragonal iron matrix, and as the carbon concentration can be greatly in excess of the equilibrium concentration in ferrite, the strength is raised very substantially. High carbon martensites are normally very hard but brittle, the yield strength reaching as much as 1500 MN m^{-2}; much of this increase can be directly attributed to increased interstitial solid solution hardening, but there is also a contribution from the high dislocation density which is characteristic of martensitic transformations in iron-carbon alloys. Martensite will be dealt with in more detail in Chapter 5, which shows that by subjecting it to a further heat treatment at intermediate temperatures (*tempering*), a proportion of the strength is retained, with a substantial gain in the toughness and ductility of the steel.

2.4 Substitutional solid solution strengthening of iron

Many metallic elements form solid solutions in γ- and α-iron. These are invariably substitutional solid solutions, but for a constant atomic concentration of alloying elements there are large variations in strength. Using single crystal data for several metals, Fig. 2.6 shows that an element such as vanadium has a weak strengthening effect on α-iron at low concentrations (< 2 at %), while silicon and molybdenum are much more effective strengtheners. Other data indicates that phosphorus, manganese,

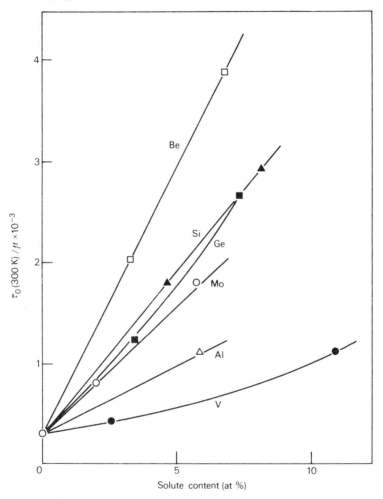

Fig. 2.6 Solid solution strengthening of iron crystals by substitutional solutes. Ratio of the critical resolved shear strees τ_0 to shear modulus μ as a function of atomic concentration (Takeuchi, J. Phys. Soc. Japan, 1969, **27**, 929)

nickel and copper are also effective strengtheners. However, it should be noted that the relative strengthening may alter with the temperature of testing, and with the concentrations of interstitial solutes present in the steels.

The strengthening achieved by substitutional solute atoms is, in general, greater the larger the difference in atomic size of the solute from that of iron, applying the Hume-Rothery size effect. However, from the work of Fleischer and Takeuchi it is apparent that differences in the elastic behaviour of solute and solvent atoms are also important in determining the overall strengthening achieved. In practical terms, the contribution to strength from solid solution effects is superimposed on hardening from other sources, e.g. grain size and dispersions. Also it is a strengthening increment, like that due to grain size, which need not adversely affect ductility. In industrial steels, solid solution strengthening is a far from negligible factor in the overall strength, where it is achieved by a number of familiar alloying elements, e.g. manganese, silicon, nickel, molybdenum, several of which are frequently present in a particular steel and are additive in their effect. These alloying elements are usually added for other reasons, e.g. Si to achieve deoxidation, Mn to combine with sulphur or Mo to promote hardenability. Therefore, the solid solution hardening contribution can be viewed as a useful bonus.

2.5 Grain size

The refinement of the grain size of ferrite provides one of the most important strengthening routes in the heat treatment of steels. The first scientific analysis of the relationship between grain size and strength, carried out on ARMCO iron by Hall and Petch, led to the Hall-Petch relationship between the yield stress σ_y and the grain diameter d,

$$\sigma_y = \sigma_0 + k_y d^{-\frac{1}{2}} \tag{2.9}$$

where σ_0 and k_y are constants. This type of relationship holds for a wide variety of irons and steels as well as for many non-ferrous metals and alloys. A typical set of results for mild steel is given in Fig. 2.7, where the linear relationship between σ_y and $d^{-\frac{1}{2}}$ is clearly shown for three test temperatures.

The constant σ_0 is called the *friction stress*. It is the intercept on the stress axis, representing the stress required to move free dislocations along the slip planes in the bcc crystals, and can be regarded as the yield stress of a single crystal ($d^{-\frac{1}{2}} = 0$). This stress is particularly sensitive to temperature (Fig. 2.7) and composition. The k_y term represents the slope of the σ_y–$d^{-\frac{1}{2}}$ plot which has been found not to be sensitive to temperature (Fig. 2.7), composition and strain rate.

In line with the Cottrell-Bilby theory of the yield point involving the break away of dislocations from interstitial carbon atmospheres, k_y has been

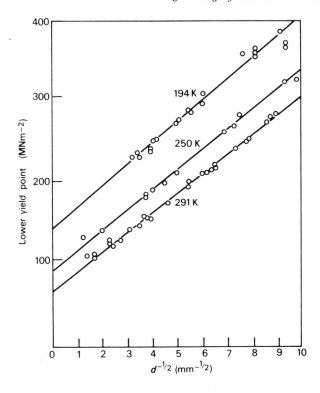

Fig. 2.7 Dependence of the lower yield stress of mild steel on grain size (Petch, In: *Fracture*, ed. Averbach *et al.*, John Wiley, 1959)

referred to as the unpinning parameter. However, the insensitivity of k_y to temperature suggests that unpinning rarely occurs, and emphasizes the theory that new dislocations are generated at the yield point. This is consistent with the theories explaining the yield point in terms of the movement of new dislocations, the velocities of which are stress dependent (Section 2.3.2).

The grain size effect on the yield stress can therefore be explained by assuming that a dislocation source operates within a crystal causing dislocations to move and eventually to pile up at the grain boundary. The pile-up causes a stress to be generated in the adjacent grain, which, when it reaches a critical value, operates a new source in that grain. In this way, the yielding process is propagated from grain to grain. This can be observed macroscopically by the passage of a Luders band. The grain size determines the distance dislocations have to move to form grain boundary pile-ups, and thus the number of dislocations involved. With large grain sizes, the pile-ups will contain larger numbers of dislocations which will in turn cause higher stress concentrations in neighbouring grains. The shear stress τ_i at the head of a dislocation pile-up is equal to $n\tau$ where n is the number of dislocations involved, and τ is the shear stress on the slip plane. This means that the coarser the grain size, the easier it will be to propagate the yielding process.

In practical terms, the finer the grain size, the higher the resulting yield stress and, as a result, in modern steel working much attention is paid to the final ferrite grain size. While a coarse grain size of $d^{-\frac{1}{2}} = 2$, i.e. $d = 0.25$ mm, gives a yield stress in mild steels of around 100 MN m^{-2}, grain refinement to $d^{-\frac{1}{2}} = 20$, i.e. $d = 0.0025$ mm, raises the yield stress to over 500 MN m^{-2}, so that achieving grain sizes in the range 2–10 μm is extremely worthwhile. Over the last 20 years, developments in rolling practice and the addition of small concentrations of particular alloying elements to mild steels, have resulted in dramatic improvements in the mechanical properties of this widely-used engineering material (Chapter 9).

2.6 Dispersion strengthening

In all steels there is normally more than one phase present, and indeed it is often the case that several phases can be recognised in the microstructure. The matrix, which is usually ferrite (bcc structure) or austenite (fcc structure) strengthened by grain size refinement and by solid solution additions, is further strengthened, often to a considerable degree, by controlling the dispersions of the other phases in the microstructure. The commonest other phases are carbides formed as a result of the low solubility of carbon in α-iron. In plain carbon steels this carbide is normally Fe_3C (cementite) which can occur in a wide range of structures from coarse lamellar form (pearlite), to fine rod or spheroidal precipitates (tempered steels). In alloy steels, the same range of structures is encountered, except that in many cases iron carbide is replaced by other carbides which are thermodynamically more stable. Other dispersed phases which are encountered include nitrides, intermetallic compounds, and, in cast irons, graphite.

Most dispersions lead to strengthening, but often they can have adverse effects on ductility and toughness. In fine dispersions, ideally small spheres randomly dispersed in a matrix, there are well-defined relationships between the yield stress, or initial flow stress, and the parameters of the dispersion. The simplest is that due to Orowan relating the yield stress of the dispersed alloy τ_0 to the interparticle spacing Λ:

$$\tau_0 = \tau_s + \frac{T}{b\Lambda/2} \qquad (2.10)$$

where τ_s is the yield strength of matrix, T is the line tension of dislocation and b is Burgers vector. This result emerges from an analysis of the movement of dislocations around spherical particles, showing that the yield stress varies inversely as the spacing between the particles. If the dispersion is coarsened by further heat treatment, the strength of the alloy falls. A more precise form of the Orowan equation, due to Ashby, takes into account the radius r of the particles:

$$\tau_0 = \tau_s + \frac{Gb}{4r} \phi \ln\left(\frac{\Lambda - 2r}{2b}\right)\left(\frac{1}{(\Lambda - 2r)/2}\right) \quad (2.11)$$

where ϕ is a constant and G is the shear modulus.

These relationships can be applied to simple dispersions sometimes found in steels, particularly after tempering, when, in plain carbon steels, the structure consists of spheroidal cementite particles in a ferritic matrix. However, they can provide approximations in less ideal cases, which are the rule in steels, where the dispersions vary over the range from fine rods and plates to irregular polyhedra.

Perhaps the most familiar structure in steels is that of the eutectoid pearlite, usually a lamellar mixture of ferrite and cementite. This can be considered as an extreme form of dispersion of one phase in another, and undoubtedly provides a useful contribution to strengthening. The lamellar spacing can be varied over wide limits, and again the strength is sensitive to such changes (see chapter 3). When the coarseness of the pearlite is represented by a mean uninterrupted free ferrite path (MFFP) in the pearlitic ferrite, it has been shown that the flow stress is related to $\text{MFFP}^{-\frac{1}{2}}$, i.e. there is a relationship of the Hall-Petch type (Fig. 2.8).

2.7 An overall view

Strength in steels arises from several phenomena, which usually contribute collectively to the observed mechanical properties. The heat treatment of steels is aimed at adjusting these contributions so that the required balance of mechanical properties is achieved. Fortunately the γ/α phase change allows great variations in microstructure to be produced, so that a wide range of mechanical properties can be obtained even in plain carbon steels. The additional use of metallic alloying elements, primarily as a result of their influence on the transformation, provides an even greater control over microstructure, with consequent benefits in the mechanical properties.

2.8 Some practical aspects

The presence of a sharp yield point in a steel can be detrimental to its behaviour, for example, when used for pressings, where complex patterns of Luders bands can produce rough surfaces and lead to poor workability. The severity of the yield point is directly related to the amount of carbon and nitrogen in solution in the ferrite, so that steps taken to reduce these concentrations are helpful. Unfortunately, yield points can be obtained

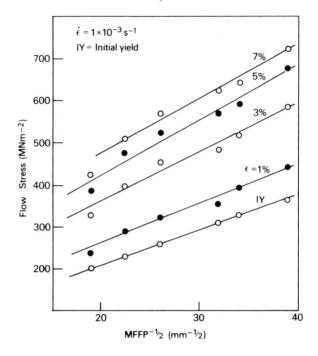

Fig. 2.8 Dependence of the flow stress at several strains on the mean free ferrite path in a pearlitic steel (Takaheshi and Nagumo, *Trans. Japan Institute of Metals*, 1970, **11**, 113)

with very low concentrations of carbon and nitrogen, and it is impracticable in industrial conditions to obtain steels below these limits. However, any heat treatment which reduces interstitial solid solution is beneficial, e.g. slow cooling from annealing treatments. The yield point can be more reliably eliminated prior to working by a small amount of cold rolling (0.5–2%), referred to as *temper rolling*. As both nitrogen and carbon diffuse appreciably in ferrite at ambient temperatures, it is desirable to fabricate steels soon after rolling and annealing.

While carbon and nitrogen can both cause strain ageing and consequently a yield point, the higher solubility of nitrogen in ferrite means that it provides the greater problem in steels used for deep drawing and pressing. Steps are taken during steelmaking to keep the nitrogen level down, but to minimize its effects, the easiest solution is to add small concentrations of strong nitride formers such as aluminium, titanium or vanadium, which reduce the nitrogen in solution to very low levels.

The occurrence of strain ageing can, by increasing both the yield stress and ultimate tensile stress, benefit mild steels which are used for constructional purposes. Furthermore, the fatigue properties are improved, both at room temperature and in the range up to 350°C. The existence of a well-defined fatigue limit in steels, i.e. a fatigue stress limit below which failure does not occur, has been linked to the occurrence of strain ageing during the test, but even very pure iron shows the same behaviour. It should be emphasized that even in a relatively simple low carbon steel, the strength arises not only from these effects of carbon and nitrogen, but also from the solid solution hardening of elements such as silicon and manganese, and potentially from the refinement of the ferrite grain size by various means.

Further reading

Cottrell, A. H., *Dislocations and Plastic Flow in Crystals*, Oxford University Press, 1953

Baird, J. D., *Strain Ageing of Steel—A Critical Review*, reprinted from Iron and Steel, 1963

Metallurgical Developments in Carbon Steels, Iron and Steel Institute Special Report, **81**, 1963

AIME, *Iron and its Dilute Solutions*, Interscience, New York, 1963

Speich, G. R. and Clark, J. B. (eds), *Precipitation from Iron-base Alloys*, Gordon and Breach, 1965

Langer, E. W., *The Quench Ageing Process in Iron*, Copenhagen, 1967

Honeycombe, R. W. K., *The Plastic Deformation of Metals*, Edward Arnold, 1968

Takeuchi, S., *J. Phys. Soc. Japan* **27**, 929, 1969

Leslie, W. C., *Met. Trans.* **3**, 5, 1972

Hall, E. O., *Yield Point Phenomena in Metals and Alloys*, Macmillan, 1970

Pickering, F. B., *Physical Metallurgy and the Design of Steels*, Applied Science Publishers, London, 1978

3

The iron-carbon equilibrium diagram and plain carbon steels

3.1 The iron-carbon equilibrium diagram

A study of the constitution and structure of all steels and irons must first start with the iron-carbon equilibrium diagram. Many of the basic features of this system (Fig. 3.1) influence the behaviour of even the most complex alloy steels. For example, the phases found in the simple binary Fe-C system persist in complex steels, but it is necessary to examine the effects alloying elements have on the formation and properties of these phases. The iron-carbon diagram provides a valuable foundation on which to build knowledge of both plain carbon and alloy steels in their immense variety.

It should first be pointed out that the normal equilibrium diagram really represents the metastable equilibrium between iron and iron carbide (cementite). Cementite is metastable, and the true equilibrium should be between iron and graphite. Although graphite occurs extensively in cast irons (2–4 wt% C), it is usually difficult to obtain this equilibrium phase in steels (0.03–1.5 wt% C). Therefore, the metastable equilibrium between iron and iron carbide should be considered, because it is relevant to the behaviour of most steels in practice.

The much larger phase field of γ-iron (austenite) compared with that of α-iron (ferrite) reflects the much greater solubility of carbon in γ-iron, with a maximum value of just over 2 wt% at 1147°C (E, Fig. 3.1). This high solubility of carbon in γ-iron is of extreme importance in heat treatment, when solution treatment in the γ-region followed by rapid quenching to room temperature allows a supersaturated solid solution of carbon in iron to be formed. The α-iron phase field is severely restricted, with a maximum carbon solubility of 0.02 wt% at 723°C (P), so over the carbon range encountered in steels from 0.05 to 1.5 wt%, α-iron is normally associated with iron carbide in one form or another. Similarly, the δ-phase field is very restricted between 1390 and 1534°C and disappears completely when the carbon content reaches 0.5 wt% (B).

There are several temperatures or critical points in the diagram which are important, both from the basic and from the practical point of view. Firstly, there is the A_1 temperature at which the eutectoid reaction occurs (P-S-K), which is 723°C in the binary diagram. Secondly, there is the A_3 temperature when α-iron transforms to γ-iron. For pure iron this occurs at 910°C, but the transformation temperature is progressively lowered along the line GS

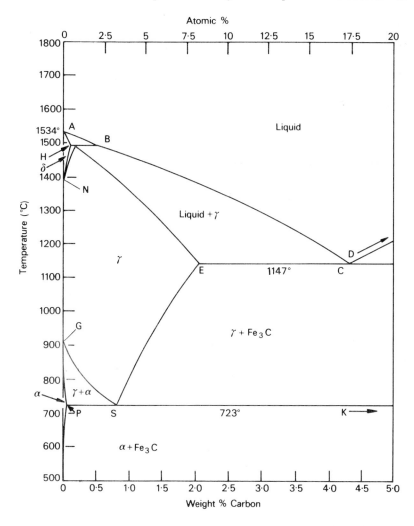

Fig. 3.1 The iron-carbon diagram (after Hansen, *Constitution of Binary Alloys*, 2nd ed, McGraw Hill, 1958)

by the addition of carbon. The third point is A_4 at which γ-iron transforms to δ-iron, 1390°C in pure iron, but this is raised as carbon is added. The A_2 point is the Curie point when iron changes from the ferro- to the paramagnetic condition. This temperature is 769°C for pure iron, but no change in crystal structure is involved. The A_1, A_3, and A_4 points are easily detected by thermal analysis or dilatometry during cooling or heating cycles, and some hysteresis is observed. Consequently, three values for each point can be obtained, Ac for heating (chauffage), Ar for cooling (refroidissement) and Ae (equilibrium), but it should be emphasized that the

Ac and Ar values will be sensitive to the rates of heating and cooling, as well as to the presence of alloying elements.

The great difference in carbon solubility between γ- and α-iron leads normally to the rejection of carbon as iron carbide at the boundaries of the γ-phase field. The transformation of $\gamma \to \alpha$-iron occurs via a eutectoid reaction which plays a dominant role in heat treatment. The eutectoid temperature is 723°C while the eutectoid composition is 0.80% C(s). On cooling alloys containing less than 0.80% C slowly, hypo-eutectoid ferrite is formed from austenite in the range 910–723°C with enrichment of the residual austenite in carbon, until at 723°C the remaining austenite, now containing 0.8% carbon transforms to pearlite, a lamellar mixture of ferrite and iron carbide (cementite) (Fig. 3.2a). In. austenite with 0.80 to 2.06% carbon, on cooling slowly in the temperature interval 1147°C to 723°C, cementite first forms progressively depleting the austenite in carbon, until at 723°C, the austenite contains 0.8% carbon and transforms to pearlite.

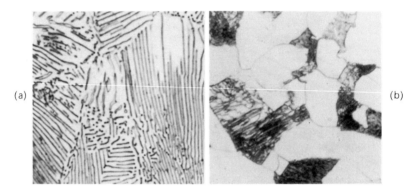

Fig. 3.2 a, 0.8C steel-pearlite (Ricks). Optical micrograph, ×1000; b, 0.4C steel–ferrite and pearlite (Ricks). Optical micrograph, ×1100

Steels with less than about 0.8% carbon are thus hypo-eutectoid alloys with ferrite and pearlite as the prime constituents (Fig. 3.2b), the relative volume fractions being determined by the lever rule which states that as the carbon content is increased, the volume percentage of pearlite increases, until it is 100% at the eutectoid composition. Above 0.8% C, cementite becomes the hyper-eutectoid phase, and a similar variation in volume fraction of cementite and pearlite occurs on this side of the eutectoid composition.

The three phases, ferrite, cementite and pearlite are thus the principle constituents of the microstructure of plain carbon steels, provided they have been subjected to relatively slow cooling rates to avoid the formation of metastable phases. Consequently it is important to examine the nucleation and growth of these phases, and to determine the factors which control their morphology.

3.2 The austenite-ferrite transformation

Under equilibrium conditions, pro-eutectoid ferrite will form in iron-carbon alloys containing up to 0.8 % carbon. The reaction occurs at 910°C in pure iron, but takes place between 910°C and 723°C in iron-carbon alloys. However, by quenching from the austenitic state to temperatures below the eutectoid temperature Ae_1, ferrite can be formed down to temperatures as low as 600°C. There are pronounced morphological changes as the transformation temperature is lowered, which it should be emphasized apply in general to hypo- and hyper-eutectoid phases, although in each case there will be variations due to the precise crystallography of the phases involved. For example, the same principles apply to the formation of cementite from austenite, but it is not difficult to distinguish ferrite from cementite morphologically.

As a result of a survey of the behaviour of plain carbon steels, Dubé proposed a classification of morphologies of ferrite which occur as the γ/α transformation temperature is lowered. Dubé recognized four well-defined morphologies, later extended by Aaronson:

(1) Grain boundary allotriomorphs: These are crystals which nucleate at the austenite grain boundaries, and at the highest temperatures (800–850°C) have curved boundaries with the austenite (Fig. 3.3a). They are usually equi-axed or lenticular in form. As the temperature of transformation is lowered, these crystals develop facets on at least one side, but often on both sides of the boundary.

(2) Widmanstätten side plates or laths: These plates nucleate at the γ-grain boundaries, but grow along well-defined matrix planes. They grow either direct from the boundaries, or nucleate on pre-existing ferrite allotriomorphs (Fig. 3.3b).

(3) Intragranular idiomorphs: These comprise roughly equi-axed crystals which nucleate within the austenitic grains (Fig. 3.3c), and possess either irregular curved boundaries, or boundaries with better-defined crystallographic characteristics.

(4) Intragranular plates: These plates are similar to those growing from the grain boundaries, but they nucleate entirely within the austenite grains (Fig. 3.3d).

Grain boundary allotriomorphs are the first morphology to appear over the whole range of composition and temperature. However, at the highest temperatures (above 800°C), they predominate by growing along the boundaries, and also into the grains to give a well-defined grain structure, generally referred to as equi-axed ferrite. The allotriomorphs nucleate having a Kurdjumov-Sachs orientation relationship with one austenite grain (γ_1):

$$\{111\}_{\gamma_1} // \{110\}_\alpha$$
$$\langle 110 \rangle_{\gamma_1} // \langle 111 \rangle_\alpha$$

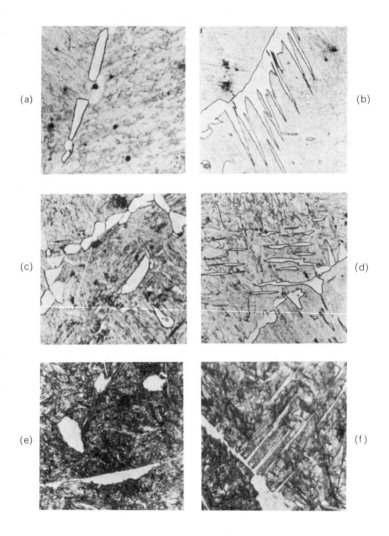

Fig. 3.3 Growth of pro-eutectoid ferrite and hypereutectoid cementite: a, 0.34C steel, 12 min at 790°C. Grain boundary allotriomorphs of ferrite; b, 0.34C steel, 15 min at 725°C. Widmanstätten ferrite growing from grain boundary ferrite; c, 0.34C steel, 12 min at 790°C. Grain boundary allotriomorphs and intragranular idiomorphs of ferrite; d, 0.34C steel, 15 min at 725°C. Intragranular Widmanstätten ferrite plates; e, 1.2C steel, 10 min at 730°C. Grain boundary allotriomorphs and intragranular idiomorphs of cementite; f, 1.2C steel, 10 min at 730°C. Widmanstätten cementite; Optical micrographs, a–d ×500, e and f ×350 (courtesy of R. A. Ricks)

But they also grow into the adjacent austenite grain (γ_2) with which they should normally have a random orientation relationship. The disordered boundary responsible for this growth should migrate more readily at high temperatures, although there is evidence that such boundaries readily develop growth facets, indicating that there is anisotropy of growth rate.

At lower transformation temperatures, the mobility of curved or random γ/α boundaries decreases, while the coherent interfaces become more dominant. For example, laths (narrow plates) of ferrite grow from protuberances on the grain boundary ferrite on the side of the coherent boundary, so the laths are moving into austenite with which they have the Kurdjumov-Sachs relationship. The laths can also grow from clean austenite grain boundaries, the net result being a structure which is normally referred to, rather inaccurately, as acicular, or, more appropriately, as Widmanstätten ferrite. This structure is encouraged by large austenite grain sizes which prevent the impingement of grain boundary ferrite by growth across grains, thus allowing Widmanstätten ferrite room to grow. If the carbon content is too high ($> 0.4\%$), the pearlitic regions are sufficiently large to prevent ferrite laths growing. However, if the carbon content is below 0.2%, impingement of allotriomorphs across γ-grains again minimizes the growth of Widmanstätten ferrite. But the most important factor is the temperature of growth of the ferrite, which is determined by the overall rate of cooling of the steel, or the temperature of isothermal transformation. Two important structural features are found in Widmanstätten ferrite:
(1) Sideways growth occurs by movement of small (incoherent) steps along the planar interfaces
(2) The formation of laths is often accompanied by surface relief effects.

In common with many other phases presenting a planar coherent or semicoherent interface to a matrix, Widmanstätten ferrite plates have been shown to grow by the lateral movement of small incoherent steps on the interface (Fig. 3.4). Some evidence has been obtained to support the view that shear displacements can be involved at low transformation temperatures. The dislocation density of the ferrite increases with decreasing transformation temperature, and surface relief is often observed after transformation. However, such effects can be obtained during diffusion-controlled reactions, and thus do not prove conclusively that a shear transformation has taken place.

3.3. The austenite-cementite transformation

The Dube classification applies equally well to the various morphologies of cementite formed at progressively lower transformation temperatures. The initial development of grain boundary allotriomorphs is very similar to that of ferrite, and the growth of side plates or Widmanstätten cementite follows the same pattern (Figs. 3.3e and f). The cementite plates are more

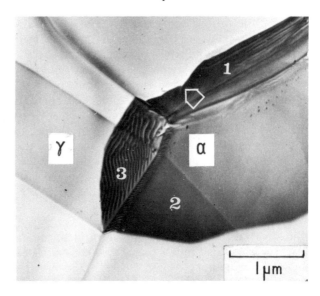

Fig. 3.4 Ferrite forming in austenite with planar and stepped interfaces in a low carbon alloy steel (Edmonds and Honeycombe). Photoemission electron micrograph

rigorously crystallographic in form, despite the fact that the orientation relationship with austenite (found by Pitsch) is a more complex one, i.e.,

$(100)_c // (5\bar{5}4)_\gamma$

$(010)_c // (110)_\gamma$

$(001)_c // (\bar{2}25)_\gamma$

As in the case of ferrite, most of the side plates originate from grain boundary allotriomorphs, but in the cementite reaction more side plates nucleate at twin boundaries in austenite.

3.4 The kinetics of the γ/α transformation

The transformation of austenite in steels can be studied during continuous cooling using various physical measurements, e.g. dilatometry, thermal analysis, electrical resistivity, etc., however the results obtained are very sensitive to the cooling rate used. Davenport and Bain first introduced the isothermal transformation approach, and showed that by studying the reaction isothermally at a series of temperatures below the Ae_1, a characteristic time-temperature-transformation or *TTT* curve can be obtained for each particular steel. In their simplest form, these transformation curves have a well-defined 'C' shape (Fig. 3.5), where the nose of the

The iron-carbon equilibrium diagram and plain carbon steels 35

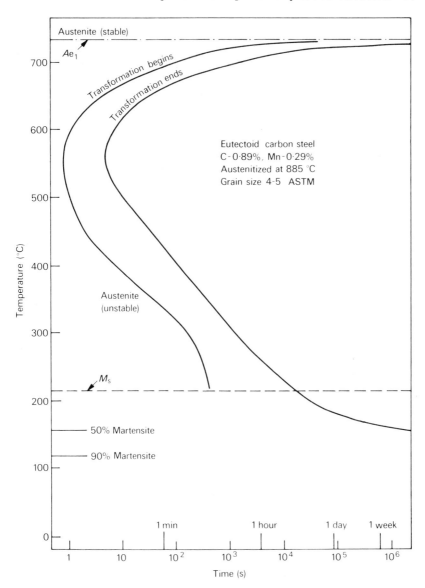

Fig. 3.5 Time-temperature-transformation (*TTT*) diagram for a 0.89 carbon steel (US Steel Co., *Atlas of Isothermal Diagrams*)

curve represents the temperature at which the reaction proceeds most rapidly, slowing down both at higher and at lower temperatures. This can be explained in general terms as follows. Near the eutectoid temperature, the degree of undercooling, ΔT, is low so the driving force for the transformation is small. However, as ΔT increases the driving force also increases, and the reaction occurs more quickly, until the maximum rate at the nose of the curve. Below this temperature, the driving force for the reaction continues to increase, but the reaction is now impeded by the lack of diffusivity of the rate controlling element, which in plain carbon steels will be carbon.

One of the simplest examples of a TTT curve is that for a 0.8% eutectoid carbon steel. In Fig. 3.5 the beginning and end of transformation over a wide temperature range is plotted to produce two curves making up the diagram. When the carbon content of the steel is lowered, the ferrite reaction will also take place and this is represented by another curve which is frequently imposed on the same diagram, and which normally precedes the pearlite reaction. Similarly, the cementite reaction can be recorded in hyper-eutectoid steels. The TTT curve strictly applies to the nucleation and growth of one phase in austenite, but at the lower transformation temperatures other constituents can appear, e.g. bainite, martensite. These have quite different characteristics to ferrite and pearlite, so they will be dealt with separately (Chapters 5 and 6).

3.4.1 Growth kinetics of ferrite

The growth of grain boundary allotriomorphs of ferrite has been studied in two directions; along the grain boundaries (lengthening) and into the austenite grains (thickening). For example, Kinsman and Aaronson have made direct measurements of thickening kinetics of ferrite in an iron-0.11% C alloy in a thermionic emission microscope. The thickening process was found to be parabolic:

$$S = \alpha t^{\frac{1}{2}} \tag{3.1}$$

where S = half thickness, and t = growth time.

Zener and later Hillert, investigated theoretically the growth of a plate with curved ends (Fig. 3.6) assuming that the transfer of carbon atoms from ferrite to austenite is controlled by the diffusivity of carbon in austenite, D_γ,

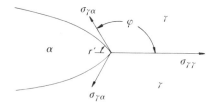

Fig. 3.6 Interfacial energies at the advancing edge of a ferrite allotriomorph (after Hillert, *Jernkontorets. Ann.*, 1957, **141**, 757)

and by the radius of curvature r' of the plate, the diffusion of carbon from the interface being more rapid the smaller the value of r'. They derived an equation, the Zener-Hillert equation, which expresses the rate of growth of a phase (ferrite) in terms of the diffusivity of the active species (carbon) in the matrix (austenite), and the relevant molar fractions of the active atomic species (carbon) in the phases. The form of the equation relevant to the lengthening of ferrite allotriomorphs is

$$G_L = \frac{D_\gamma (x_\gamma^{\gamma\alpha} - x_\gamma)}{4r'(x_\gamma - x_\alpha^{\alpha\gamma})\sin\Phi} \qquad (3.2)$$

where $x_\gamma^{\gamma\alpha}$ = molar fraction of carbon in austenite at the $\gamma/\alpha+\gamma$ phase boundary (Fig. 3.1, line GS)
x_γ = molar fraction of carbon in parent austenite
$x_\alpha^{\alpha\gamma}$ = molar fraction of carbon in ferrite at the $\alpha/\alpha+\gamma$ phase boundary (Fig. 3.1, line GP)
r' = radius of curvature of the allotriomorph adjacent to the grain junction (Fig. 3.6)
Φ = equilibrium growth angle determined by the relative energies of the interphase and grain boundaries (Fig. 3.6).

The $\sin\Phi$ term is present because each side of the allotriomorph makes an angle $\Phi - \pi/2$ with the γ/γ grain boundary (Fig. 3.6).

To apply the equation to the growth of ferrite side plates (Widmanstätten ferrite) it is simply necessary to replace the term $r'\sin\Phi$ by r, the radius of curvature of the edge of the plate. More precise expressions for the growth rate G can be obtained by allowing for the variation of D_γ with carbon concentration and temperature.

Thickening studies on Widmanstätten ferrite plates using thermionic emission microscopy have shown that the growth does not occur smoothly, but in a series of steps (Fig. 3.4). This explains the irregularities observed on thickness/time plots (Fig. 3.7). Similar results have been obtained in other alloy systems, where it has been shown that growth does in fact occur by repeated migration of small ledges along the plate boundaries. Kinsman was able to detect coarse ledges on the broad faces of ferrite plates, the height of which varied from 4×10^{-6} to 1.6×10^{-4} cm. The velocity of movement of such ledges appeared to be constant, but some irregularities were observed. Jones and Trivedi found that the velocity v was determined by:

$$v = \frac{D_\gamma(x_\gamma^{\gamma\alpha} - x_\gamma)}{a\beta(x_\gamma^{\gamma\alpha} - x_\alpha^{\alpha\gamma})} \qquad (3.3)$$

where the x terms are the relevant molar fractions, a is the ledge height and the constant β is a function of a velocity parameter $p = \dfrac{va}{2D}$. There was good

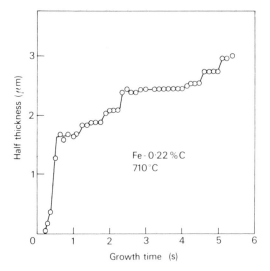

Fig. 3.7 Growth of ferrite plates in an Fe-0.22C alloy at 710°C (Aaronson et al., In: *Phase Transformations*, ASM, 1970)

agreement between observed and calculated velocities, so it is again concluded that the rate-controlling process is carbon diffusion in austenite.

3.5 The austenite-pearlite reaction

Pearlite is probably the most familiar microstructural feature in the whole science of metallography (Figs. 3.8a and b). It was discovered by Sorby over 100 years ago, who correctly assumed it to be a lamellar mixture of iron and

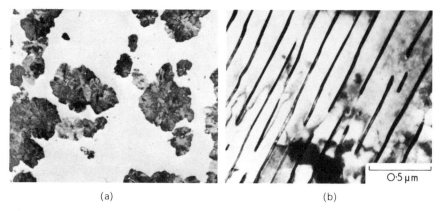

Fig. 3.8 Isothermal transformation of a 0.8C steel, 10 s at 650°C; a, optical micrograph, ×80; b, thin-foil EM of part of a pearlite nodule, ×34K (Ohmori)

The iron-carbon equilibrium diagram and plain carbon steels

iron carbide. Pearlite is a very common constituent of a wide variety of steels, where it provides a substantial contribution to strength, so it is not surprising that this phase has received intensive study. Lamellar eutectoid structures of this type are widespread in metallurgy, and frequently pearlite is used as a generic term to describe them. These structures have much in common with the cellular precipitation reactions. Both types of reaction occur by nucleation and growth (Fig. 3.8a), and are, therefore, diffusion controlled. Pearlite nuclei occur on austenite grain boundaries, but it is clear that they can also be associated with both pro-eutectoid ferrite and cementite. In commercial steels, pearlite nodules can nucleate on inclusions.

3.5.1 The morphology of pearlite

The idealized view of pearlite is a hemispherical nodule nucleated at an austenite grain boundary, and growing gradually into one austenite grain (Fig. 3.9). Apart from examining possible sites for nucleation, the following information is needed:
(a) how the lamellae increase in number
(b) the crystallographic relationships between the phases
(c) the nature of the pearlite/austenite interface
(d) the rate controlling process.

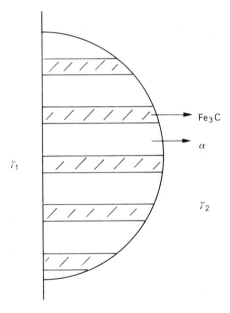

Fig. 3.9 Idealized pearlite nodule at austenite grain boundary

Not all these questions can yet be fully answered, but there is a fairly clear picture emerging. Following the classical work of Mehl and colleagues, Hillert and co-workers were able to show that pearlite could be nucleated either by ferrite, or by cementite, depending on whether the steel was hypo- or hyper-eutectoid in composition. They came to this conclusion after observing lattice continuity between the ferrite in pearlite and pro-eutectoid ferrite, as well as between cementite in pearlite and hyper-eutectoid cementite.

Mehl and co-workers took the view that pearlite nodules formed by sideways nucleation and edge-ways growth (Fig. 3.9). In this way, the rapid increase in the number of lamellae in a nodule which occurred during growth could be explained, but Modin indicated that this could equally well result from the branching of lamellar during growth. More recent thin-foil electron microscopy work by Dippenaar and Honeycombe on 12%Mn 0.8% carbon steel allowed the examination of very small nodules at an early stage of growth in an austenitic matrix rendered stable by addition of manganese. This steel is hyper-eutectoid, so grain boundary cementite forms prior to nucleation of pearlite which frequently takes place on the cementite. This work showed conclusively the continuity of grain boundary and pearlitic cementite (Fig. 3.10a), and also indicated that both the cementite and ferrite possessed unique orientations within a particular nodule. Fig. 3.10a also shows the beginning of branching of the Fe_3C lamella. However, in other nodules, sideways nucleation of laths of cementite and ferrite was observed. Nucleation of pearlite also took place on clean austenite boundaries. Hillert has shown that nucleation also occurs on ferrite, so all three types of site are effective, and the predominant sites will be determined primarily by the composition.

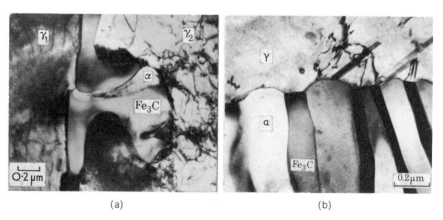

(a)   (b)

Fig. 3.10 Fe-13Mn-0.8C partly transformed at 600°C. Austenite is retained in conjunction with ferrite and cementite: a. nucleation of a pearlite nodule on grain boundary cementite; b. interface of nodule with austenite. Thin-foil EMs (Dippenaar)

C. S. Smith first pointed out that the moving pearlite interface in contact with austenite was an incoherent high energy interface growing into a grain with which the pearlitic ferrite and cementite had no orientation relationship. Therefore, the nodules which nucleated on pre-existing grain boundary cementite and ferrite would choose the higher energy interfaces across which the boundary phase had no orientation relationship with the adjacent austenite. Hillert and co-workers were able to show by suitable heat treatments that pearlite did nucleate in this way, while on the low energy interfaces Widmanstätten growth of ferrite (or cementite) was usually observed. Electron microscopy observations have confirmed that the pearlite interface with austenite is an incoherent one. Fig. 3.10b shows a typical interface on a 12%Mn 0.8% steel, where the untransformed austenite has been retained at room temperature.

The spacing of the lamellae in pearlite is a sensitive parameter which, in a particular steel, is larger the higher the transformation temperature. The spacing was first measured systematically for a number of steels by Mehl and co-workers, who demonstrated that the spacing decreased as the degree of undercooling, ΔT, below Ae_1 increased. Zener provided the first theoretical analysis of these observations by considering a volume of pearlite (Fig. 3.11) of depth δ and interlamellar spacing S_0 growing unidirectionally in the x-direction. If growth is allowed to occur by dx then the volume of austenite transformed per lamellar spacing is $S_0 \delta\, dx\, \rho$, where ρ is the density. The free energy G, available to form this volume of pearlite is

$$G = \Delta H \left(\frac{T_e - T}{T_e} \right) S_0 \delta\, dx\, \rho \qquad (3.4)$$

where $T_e = Ae_1$
T = transformation temperature
ΔH = latent heat of transformation.

The formation of this new volume of pearlite causes an increase in interfacial energy by virtue of the new ferrite and cementite interfaces formed. Therefore,

increase in interfacial area = $2\delta\, dx$

and

increase in interface energy = $2\delta\, dx\, \gamma$ \qquad (3.5)

where γ = interfacial energy per unit area.

Growth of the lamellae can only occur, if the increase in surface energy is less than the decrease in energy resulting from the transformation. Therefore, the condition for growth can be found from equations (3.4) and (3.5):

$$\Delta H \left(\frac{T_e - T}{T_e} \right) \rho S_0 = 2\gamma \qquad (3.6)$$

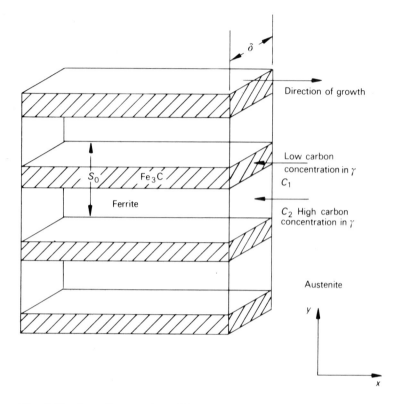

Fig. 3.11 A pearlite growth model

This is a very simple treatment which neglects any strain energy term. Also the free energy change is found from the enthalpy change per unit mass, and it assumes that the specific heats of austenite and pearlite are identical. Nevertheless, the equation predicts three important aspects of the transformation:

(1) The pearlite spacing S_0 decreases with decreasing transformation temperature

(2) The fineness of the spacing is limited by the free energy available from the transformation

(3) A linear relation should exist between the reciprocal of the spacing and the degree of undercooling.

The dependence of spacing on temperature for several plain carbon and alloy steels is shown in Fig. 3.12, where it is seen that the Zener analysis holds at lower degrees of supercooling, but as ΔT increases the results are more scattered. The earlier spacing measurements can be criticized, partly because of the difficulty of making effective measurements on nodules with complex lamellar morphology. In recent years this problem has been

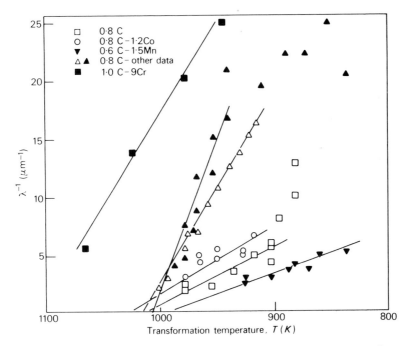

Fig. 3.12 Reciprocal of the interlamellar spacing of pearlite from several alloys as a function of temperature

eliminated by causing pearlite to grow unidirectionally by imposing a large temperature gradient along a steel rod. This technique leads to regular spacings, and closer correlation of the spacing with the velocity of growth.

3.5.2 The crystallography of pearlite

In a typical pearlite nodule there are two interpenetrating single crystals of ferrite and of cementite, neither of which is orientation-related to the austenite grain in which they are growing. However, there is always a well-defined crystallographic orientation between the cementite and ferrite lamellae within a pearlite nodule. At least two different relationships have been identified, the most important being

Pitsch/Petch relationship
  $(001)_c // (5\bar{2}1)_\alpha$
  $(010)_c\ 2-3°$ from $[11\bar{3}]_\alpha$
  $(100)_c\ 2-3°$ from $[13\bar{1}]_\alpha$
Bagaryatski relationship
  $(100)_c // (0\bar{1}1)_\alpha$
  $(010)_c // (1\bar{1}\bar{1})_\alpha$
  $(001)_c // (211)_\alpha$

The two relationships are found side by side in the same steel, and the frequency of each varies rather unpredictably. Thin-foil electron microscopy has shown that the pearlite nodules nucleating on clean austenite boundaries exhibit the Pitsch/Petch relationship. The pearlitic ferrite is related to the austenite grain γ_1, (Fig. 3.9) into which it is not growing. The relationship is always close to the Kurdjumov-Sachs relationship. Also the pearlitic cementite is related to austenite grain, γ_1, by a relationship found by Pitsch for Widmanstätten cementite in austenite. Both the pearlitic cementite and ferrite are unrelated to austenite grain, γ_2.

In contrast, the Bagaryatski relationship is found to hold for pearlite nodules nucleated on hyper-eutectoid cementite, usually formed at the austenite grain boundaries. In this case, the pearlitic cementite is related to austenite grain γ_1 by the Pitsch relationship for Widmanstätten cementite, while the pearlitic ferrite is not related to grain γ_1. Clearly the grain boundary cementite shields the newly-formed ferrite from any contact with γ_1. It also follows that the grain boundary cementite and the pearlitic cementite are continuous, i.e. of the same orientation. Again, neither the pearlitic ferrite or cementite are related to austenite grain γ_2.

It is, therefore, predicted that Pitsch/Petch-type colonies predominate as the true eutectoid composition is approached, whereas Bagaryatski-type colonies should prevail at higher carbon levels. It is also likely that the Bagaryatski relationship will become more dominant in hypo-eutectoid steels as the carbon level is reduced, but this has not yet been conclusively proved.

3.5.3 The kinetics of pearlite formation

The formation of pearlite is a good example of a nucleation and growth process. The pearlite nucleates at preferred sites in the austenite and the nuclei then grow until they impinge on each other. The process is both time and temperature-dependent, as it is controlled by the diffusivity of the relevant atoms. Johnson and Mehl first applied a detailed analysis of nucleation and growth to the pearlite reaction, which assumed that the fraction of austenite transformed (X) could be expressed in terms of a rate of nucleation $\dot{N}$ defined as the number of nuclei per unit volume of untransformed austenite formed per second, and a rate of growth of these nuclei G, expressed as radial growth in cm s^{-1}. They made certain simplifying assumptions of which the most significant were:
(1) Nucleation was regarded as a random event
(2) The rate of nucleation $\dot{N}$ was assumed to be constant with time
(3) The rate of growth G was assumed to be constant with time
(4) The nuclei were regarded as spherical and in due course impinged on neighbouring spheres.

An expression was obtained for the fraction of austenite transformed X, in time t:

$$X = 1 - e^{-\frac{\pi}{3}\dot{N}G^3 t^4} \qquad (3.7)$$

This relationship gives a sigmoidal type of curve, when X is plotted against t for chosen values of $\dot{N}$ and G. A typical curve is shown in Fig. 3.13a for particular values of $\dot{N}$ and G. If X is plotted against $\sqrt[4]{\dot{N}G^3}\,t$, a sigmoidal master curve is obtained which expresses the basic kinetic behaviour expected of a nucleation and growth process in a given alloy (Fig. 3.13b).

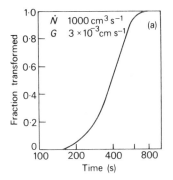
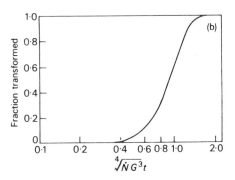

Fig. 3.13 Kinetics of pearlite reaction: a, calculated curve for specific N and G; b, master reaction curve for general nucleation (Mehl and Hagel, *Progress in Metal Physics*, 1956, **6**, 74)

In practice, however, the pearlitic reaction does not conform to the simple nucleation and growth model referred to above. Amongst the difficulties, the following are predominant:
(1) $\dot{N}$ is not constant with time
(2) G can vary from nodule to nodule and with time
(3) The nuclei are not randomly distributed
(4) The nodules are not true spheres.

This led Cahn and Hagel to a new theoretical approach which fully recognized the inhomogenous nature of nucleation in the pearlite reaction. It was pointed out that not all grain boundary nucleation sites were equivalent, that grain corners would be more effective than edges, and that edges would be better than grain surfaces. Cahn assumed that, normally, a high rate of nucleation would occur at these special sites, and that consequently site-saturation would occur at an early stage of the reaction. In these circumstances, the reaction would then be controlled by the radial growth velocity which, in the simple theory, is assumed again to be constant.

The expression for the fraction of austenite transformed assuming site saturation of grain corner sites is:

$$X = 1 - e^{-\frac{4}{3}\pi\eta G^3 t^3} \qquad (3.8)$$

where η is the number of grain corners per unit volume. In practice, site saturation sets in before 20% transformation, so the actual nucleation rate is unimportant, and does not come into the Equation (3.8). The time for completion of the reaction, t_f, is simply defined as

$$t_f = 0.5 \, d/G \tag{3.9}$$

where d = austenite grain diameter and d/G is the time taken for one nodule to absorb one grain, so the presence of only several nodules per grain will meet the above criterion for t_f. Only at small degrees of undercooling below Ae_1 will the rate of nucleation $\dot{N}$ be sufficiently low to avoid site saturation at austenite grain boundaries. In these circumstances, $\dot{N}$ would then enter into the expression for the overall reaction rate. The nucleation rate, where measured, does seem to vary with time according to the relation

$$\dot{N} = kt^n \tag{3.10}$$

where k and n are constants. However, for most experimental conditions, the rate of growth G is the dominant quantity.

The rate of growth of pearlite nuclei can be measured by reacting a series of samples for increasing times at a particular temperature. As a result of measurements on polished and etched sections, the radius of the largest pearlite area, assumed to be a projection of the first nodule to nucleate, can be plotted against time. Normally a straight line is obtained, the slope of which is G (Fig. 3.14). It has been found that G is structure insensitive, i.e. structural changes such as grain size, presence or absence of carbide particles have little effect. However, G is markedly dependent on temperature, specifically the degree of cooling ΔT below T_e, and increases with increasing degree of undercooling until the nose of the TTT curve is reached. G is also strongly influenced by the concentration of alloying elements present.

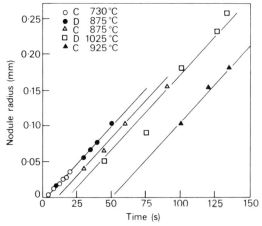

Fig. 3.14 Growth of pearlite in two 0.8C 0.6Mn steels, C and D (Mehl and Hagel, *Progress in Metal Physics*, 1956, **6**, 74)

3.5.4 The rate-controlling process

Early work on plain carbon steels assumed that the rate-controlling process in the growth of pearlite was the diffusion of carbon in austenite, and Mehl proposed the following relationship for G:

$$G = \frac{K D_c^\gamma}{S_0} \quad (3.11)$$

where D_c^γ is the diffusion coefficient of carbon in austenite, S_0 is the interlamellar spacing and K is a constant. The growth rate increases as the transformation temperature is lowered, because the driving force of the reaction is increased. However, the reaction is still diffusion-controlled so the diffusion distance must be reduced to compensate for the decrease in diffusivity with decreasing temperature. Consequently, as the temperature is lowered the pearlite interlamellar spacing is reduced.

The early theoretical treatments of Brandt and Zener, therefore, attempted to calculate the growth rate of pearlite in terms of a simple model in which the diffusion of carbon in austenite was assumed to be the rate-controlling process. Fig. 3.11 represents the model used in which a planar front of pearlite is advancing into an austenite grain. It was assumed that the carbon concentration in the austenite would be low (c_1) at the mid-points of cementite lamellae, and high at the mid-points (c_2) of ferrite lamellae. The values c_1 and c_2 were obtained from the iron-carbon equilibrium diagram using an extrapolation of the austenite-ferrite and austenite-cementite phase boundaries first proposed by Hultgren (Fig. 3.15). Brandt, by solving the applicable diffusion equation, obtained a relationship of the same form as Equation (3.11), but he was also able to evaluate K in terms of the carbon concentration differences c_1 and c_2, which are assumed to develop at the austenite-pearlite interface. Zener likewise derived an expression for G of a similar type involving two concentration terms:

$$G = \left(\frac{\Delta c}{c_p - c_\gamma}\right)\left(\frac{D_c^\gamma}{S_0}\right) \quad (3.12)$$

where c_p and c_γ are the number of solute atoms per unit volume in the two phases, and Δc is the difference in concentration in the austenite at the advancing boundary and completely away from it. This is the solute gradient which leads to diffusion.

However, these early theories have now been supplanted by others due to Hillert, Cahn and Hagel, Kirkaldy and Lundquist which have been developed to a stage where the simple iron-carbide system has been treated in a sophisticated way, and the role of alloying elements also explained. While diffusion of carbon in austenite has again been assumed to be the rate-controlling process, some treatments have assumed that boundary diffusion is rate-controlling. Hillert using this latter approach, and assuming that the austenite has periodic compositional differences along

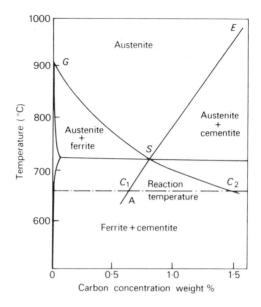

Fig. 3.15 Hultgren extrapolation of phase boundaries in Fe-C diagram (Mehl and Hagel, *Progress in Metal Physics*, 1956, **6**, 74)

the interface, depending on whether a ferrite or carbide lamella is in the vicinity, arrived at the following relationship:

$$G = \left(\frac{12 A D_b \delta S_0^2 (c_2 - c_1)}{S_\alpha S_\beta (c_\beta - c_\alpha)}\right)\left(\frac{1}{S_0^2}\right)\left(1 - \frac{S_c}{S_0}\right) \quad (3.13)$$

where D_b = interphase boundary diffusion coefficient
δ = thickness of interphase boundary
A = constant
c_1 and c_2 = concentrations previously referred to, from the Hultgren extrapolation
c_β = concentration of carbon in cementite, and
c_α = concentration of carbon in ferrite
S_0 = interlamellar spacing of the pearlite,
S_c = spacing at zero growth rate, and
S_γ and S_β = widths of the ferrite and cementite lamellae.

The equation is similar in form to equations involving volume diffusion, except that it involves an S_0^2 term rather than S_0. Also the present model must involve some volume diffusion in the austenite ahead of the interface to allow the differences in austenite composition at the interface to develop.

3.5.5 The strength of pearlite

The strength of pearlite would be expected to increase as the interlamellar spacing is decreased. Early work by Gensamer and colleagues showed that the yield stress of a eutectoid plain carbon steel, i.e. fully pearlitic, varied inversely as the logarithm of the mean free ferrite path in the pearlite. Later, Hugo and Woodhead used 3% nickel steels to obtain a uniform pearlitic structure throughout the test pieces. They confirmed that the interlamellar spacing was inversely proportional to the degree of undercooling. It was shown that both the yield strength and the ultimate tensile stress (UTS) could be linearly related to the reciprocal of the square root of the interlamellar spacing or of the degree of undercooling. Fig. 3.16 gives results for a 3%Ni 0.67%C eutectoid steel where this linear relationship is illustrated. Steels of lower carbon contents, i.e. down to 0.3%, gave similar results, when allowance was made for the presence of pro-eutectoid ferrite.

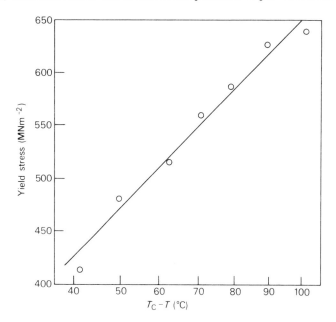

Fig. 3.16 Effect of degree of undercooling on the strength of a pearlitic nickel steel 0.67C, 0.49Mn, 2.92Ni (Hugo and Woodhead, *JISI*, 1957, **186**, 174)

The situation is rather different for lower carbon steels, i.e. below 0.3%, where pearlite occupies a substantially smaller volume of the microstructure. In these steels the yield stress is not markedly affected as the proportion of pearlite is increased, provided other factors, e.g. ferrite grain size, are kept constant. However, the tensile strength is quite sensitive to the pearlite content which is explained by the fact that there is a linear relationship between work hardening and the pearlite content (Fig. 3.17), which arises because pearlite work hardens much more rapidly than ferrite.

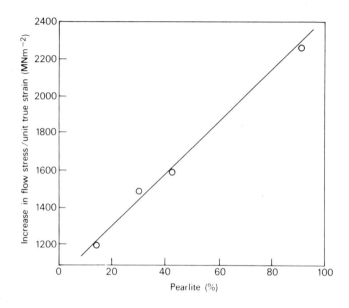

Fig. 3.17 Effect of pearlite content on work hardening (Burns and Pickering, *JISI*, 1964, **202**, 899)

Pearlite has, however, an adverse effect on ductility and toughness of plain carbon steels. For example, the impact transition temperature (see Chapter 10) is raised substantially as the carbon content is increased (Fig. 3.18), and quantitative studies have shown that 1% by volume of pearlite raises the transition temperature by about 2°C. The presence of pearlite in the microstructure provides sites of easy nucleation of cracks, particularly at the ferrite-cementite interfaces. However, as a crack can only propagate in ferrite a short distance before encountering another cementite lamella, energy is absorbed during propagation. The result is that there is a wide transition temperature range (Fig. 3.18). In contrast, the low energy absorbed overall in impact tests on pearlitic structures arises from the fact that many crack nuclei can occur at the pearlitic interfaces, together with the high work hardening rate, which restricts plastic deformation in the vicinity of the crack.

3.6 Ferrite-pearlite steels

A very high proportion of the steel used in industry has a ferrite-pearlite structure. These include a wide range of plain carbon steels where alloying

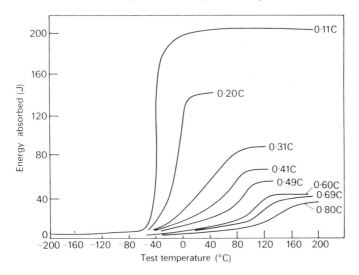

Fig. 3.18 Effect of pearlite on toughness measured by Charpy impact transition temperature (Burns and Pickering, *JISI*, 1964, **202**, 899)

additions are primarily made for steel-making purposes although they do have a strengthening role as well. For example, manganese is added to combine with sulphur, but it is also a strengthener, while both manganese and silicon are deoxidizers and aluminium is both used as a deoxidizer and as a grain refiner, and therefore a strengthener. Many low and medium alloy steels, e.g. those with nickel, give ferrite-pearlite structures, but here only essentially plain carbon steels will be dealt with.

Most plain carbon steels are not subject to heat treatment in the sense of quenching followed by tempering, but they are cooled at different rates to obtain a range of structures. Two important treatments are *normalising* and *annealing* which have special, but not very precise, meanings when applied to steels.

Normalising In the process of normalising the steel is reheated about 100°C above the Ac_3 temperature to form austenite, followed by air cooling through the phase transformation. This has as its object the refinement of the austenite and ferrite grain sizes, and the achievement of a relatively fine pearlite. It is often used after hot rolling, where a high finishing temperature can lead to a coarse microstructure.

The rate of cooling during normalizing is dependent on the dimensions of the steel, but some control can be exerted by using forced air cooling.

Annealing An annealed steel usually means one which has been austenitized at a fairly high temperature, followed by slow cooling, e.g. in a furnace. This results in transformation high in the pearlite range, giving a coarse pearlite which provides good machinability.

There are other types of annealing which are commonly practiced, e.g. *isothermal annealing*, in which the steel is cooled to a high subcritical transformation temperature, where it is allowed to transform isothermally to ferrite and coarse pearlite. *Spheroidize annealing* is applied to higher carbon pearlitic steels to improve their machinability. The steel is held at a temperature just below Ae_1 for sufficient time for the cementite lamellae of the pearlite to spheroidize. This happens because it leads to a reduction in surface energy of the cementite-ferrite interfaces.

The plain carbon ferrite-pearlite steels are essentially steels which depend for their properties on the presence of carbon and manganese. The carbon content can be varied from 0.05 wt% to 1.0 wt%, while the manganese content is from 0.25 wt% up to about 1.7 wt%. Fig. 3.19 shows the effect on the tensile strength of varying the concentration of these two elements. It has also been possible by regression analysis to determine the relative contributions to the strength of the three important mechanisms: solid solution hardening; grain size; and dispersion strengthening from lamellar pearlite. The results plotted are from steels in the normalized condition which ensures that the austenite grain sizes are roughly comparable. Variation of the carbon at constant manganese level causes a substantial increase in strength, which is almost entirely due to an increasing proportion of pearlite in the structure. The situation is rather more complex when manganese is varied at constant carbon content, as all three strengthening mechanisms are influenced. Manganese causes the eutectoid composition to occur at lower carbon contents, and so increases the proportion of pearlite in the microstructure. Manganese is also an effective solid solution strengthener, and has a grain refining influence.

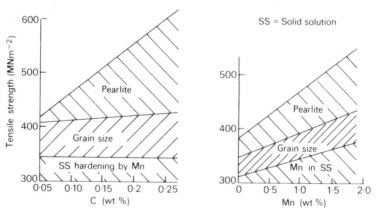

Fig. 3.19 Factors contributing to the strength of C-Mn steels (Irvine et al., JISI, 1962, **200**, 821)

It is clear that carbon provides a very cheap way of strengthening normalized steels, but the extent to which this approach can be used depends on whether the steel is to be welded or not. Welding of higher carbon steels leads to the easier formation of cracks within the weld zone, so it is usually necessary to limit the carbon content to not greater than 0.2%. In these circumstances, additional strength can then be obtained by solid solution hardening by raising the manganese content to between 1 and 1.5%.

Alternatively, refinement of the grain size can be achieved by minor alloying additions such as aluminium, vanadium, titanium and niobium, in concentrations not normally exceeding 0.1 wt% (Chapter 9). Aluminium forms a stable dispersion of AlN particles, some of which remain in the austenite grain boundaries at high temperatures, and by pinning these boundaries prevent excessive grain growth. On transformation to ferrite and pearlite, grain sizes around 12 ASTM (5–6 μm diameter) can be achieved with as little as 0.03 wt% AlN in the steel. Vanadium, titanium and niobium form very stable carbides, which also lock austenite grain boundaries, and thus allow much finer ferrite grain sizes to be achieved when the austenite transforms (Chapter 9).

Much plain carbon steel is used in the hot-finished condition, i.e. straight from hot rolling without subsequent cold rolling or heat treatment. This represents the cheapest form of steel which is usually used in low carbon and medium carbon grades, because of the loss of ductility and weldability at high carbon contents. The most important group of hot finished plain carbon steels contains less than 0.25% carbon and is used in structural shapes such as plates, I-beams, angles etc. in buildings, bridges, ships, pressure vessels and storage tanks. Hot-rolled low carbon steel sheet is an important product used extensively for fabrication where surface finish is not of prime importance. Cold rolling is used for finishing where better finish is required, and the additional strength from cold working is needed. However, for high quality sheet to be used in intricate pressing operations it is necessary to anneal the cold-worked steel to cause the ferrite to recrystallize. This is done below the Ae_1 temperature (subcritical annealing).

Carbon steels are also used extensively for closed die or drop forgings, usually in the range 0.2 to 0.5% carbon, and covering a very wide range of applications, e.g. shafts and gears. The other important field of application of plain carbon steels is as castings. Low carbon cast steels containing up to 0.25%C are widely used for miscellaneous jobbing castings as reasonable strength and ductility levels are readily obtained. Yield strengths of 240 MNm^{-2} and elongations of 30% are fairly typical for this type of steel.

Further reading

Sorby, H. C., *J. Iron Steel Inst.*, **1**, 255–288, 1887

Mehl, R. F. and Hagel, W. C., The austenite-pearlite reaction. *Prog. Met. Physics*, **6**, ed. B. Chalmers and R. King, Pergamon Press, 1956

Zackay, V. F. and Aaronson, H. I. (eds), *Decomposition of Austenite by Diffusion Processes*, Interscience, 1962

McKenzie, I. M., Recent developments in structural steels. *Met. Reviews*, **13**, 189, 1968

Kumar, R., *Physical Metallurgy of Iron and Steel*, Asia Publishing House, 1968

American Society for Metals, *Phase Transformations*, 1970

Hornbogen, E., In: *Physical Metallurgy*, ed. R. W. Cahn, North Holland Publishing Co., 2nd edition, 1970

Christian, J. W., *The Theory of Phase Transformations in Metals and Alloys*, Pergamon Press, 1965

Iron and Steel Institute, *Strong Tough Structural Steels*, Publication 104, 1967

United States Steel Corporation, *The Making, Shaping and Treating of Steel*, Chapters 40 and 41, 1971

AIME, Symposium on cellular and pearlite reactions, Detroit 1971. *Met. Trans.*, **3**, 2717–2804, 1972

Samuels, L. E., *Microscopy of Carbon Steels*, ASM, 1980

4
The effects of alloying elements on iron-carbon alloys

4.1 The γ- and α- phase fields

It would be impossible in this book to include a detailed survey of the effects of alloying elements on the iron-carbon equilibrium diagram. In the simplest version this would require analysis of a large number of ternary alloy diagrams over a wide temperature range. However, Wever pointed out that iron binary equilibrium systems fall into four main categories (Fig. 4.1): open and closed γ-field systems, and expanded and contracted γ-field systems. This approach indicates that alloying elements can influence the equilibrium diagram in two ways:

(a) by expanding the γ-field, and encouraging the formation of austenite over wider compositional limits. These elements are called γ-stabilizers.

(b) by contracting the γ-field, and encouraging the formation of ferrite over wider compositional limits. These elements are called α-stabilizers.

The form of the diagram depends to some degree on the electronic structure of the alloying elements which is reflected in their relative positions in the periodic classification.

Class 1: open γ-field To this group belongs the important steel alloying elements nickel and manganese, as well as cobalt and the inert metals ruthenium, rhodium, palladium, osmium, iridium and platinum. Both nickel and manganese, if added in sufficiently high concentration, completely eliminate the bcc α-iron phase and replace it, down to room temperature, with the γ-phase. So nickel and manganese depress the phase transformation from γ to α to lower temperatures (Fig. 4.1a), i.e. both Ae_1 and Ae_3 are lowered. It is also easier to obtain metastable austenite by quenching from the γ-region to room temperature, consequently nickel and manganese are useful elements in the formulation of austenitic steels (Chapter 11).

Class 2: expanded γ-field Carbon and nitrogen are the most important elements in this group. The γ-phase field is expanded, but its range of existence is cut short by compound formation (Fig. 4.1b). Copper, zinc and gold have a similar influence. The expansion of the γ-field by carbon, and nitrogen, underlies the whole of the heat treatment of steels, by allowing formation of a homogeneous solid solution (austenite) containing up to 2.0 wt% of carbon or 2.8 wt% of nitrogen.

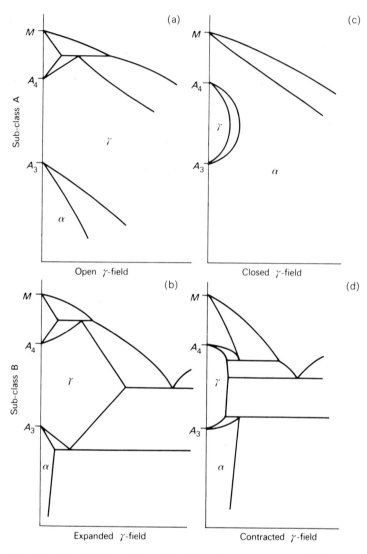

Fig. 4.1 Classification of iron alloy phase diagrams: a, open γ-field; b, expanded γ-field; c, closed γ-field; d, contracted γ-field (Wever, *Archiv. Eisenhüttenwesen*, 1928–9, **2**, 193)

Class 3: closed γ-field Many elements restrict the formation of γ-iron, causing the γ-area of the diagram to contract to a small area referred to as the gamma loop (Fig. 4.1c). This means that the relevant elements are encouraging the formation of bcc iron (ferrite), and one result is that the δ- and α-phase fields become continuous. Alloys in which this has taken place are, therefore, not amenable to the normal heat treatments involving

The effects of alloying elements on iron-carbon alloys 57

cooling through the γ/α-phase transformation. Silicon, aluminium, beryllium and phosphorus fall into this category, together with the strong carbide forming elements, titanium, vanadium, molybdenum and chromium.

Class 4: contracted γ-field Boron is the most significant element of this group, together with the carbide forming elements tantalum, niobium and zirconium. The γ-loop is strongly contracted, but is accompanied by compound formation (Fig. 4.1d).

The overall behaviour is best described in thermodynamic terms along the lines developed by Zener and by Andrews. If c_α and c_γ are the fractional concentrations of an alloying element in the α- and γ-phases, the following relation holds:

$$\frac{c_\alpha}{c_\gamma} = \beta e^{\Delta H/RT}, \text{ i.e. } \log_e \frac{c_\alpha}{c_\gamma} = \frac{\Delta H}{RT} \qquad (4.1)$$

where ΔH is the enthalpy change which is the heat absorbed per unit of solute dissolving in γ-phase minus the heat absorbed per unit of solute dissolving in α-phase, i.e. $\Delta H = H_\gamma - H_\alpha$. β is a constant.

for ferrite formers, $H_\alpha < H_\gamma$ $\therefore$ ΔH is positive
for austenite formers, $H_\alpha > H_\gamma$ $\therefore$ ΔH is negative.

In the simple treatment two fundamentally different types of equilibrium diagrams are obtained where the phase boundaries are represented by similar thermodynamic equations, but, depending on whether ΔH is positive or negative, are mirror images of each other (Fig. 4.2). In the ΔH −ve case the γ-field is unlimited, while in the ΔH +ve case, the γ-loop is introduced. ΔH will vary widely from element to element. In Fig. 4.3

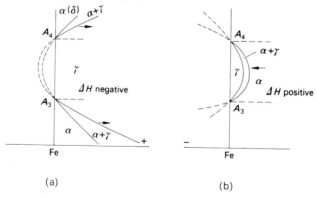

Fig. 4.2 Two basic phase diagrams: a, ΔH negative, $H_\alpha > H_\gamma$, γ favoured; b, ΔH positive, $H_\alpha < H_\gamma$, α favoured (after Zener, In: Andrews, *JISI*, 1956, **184**, 414)

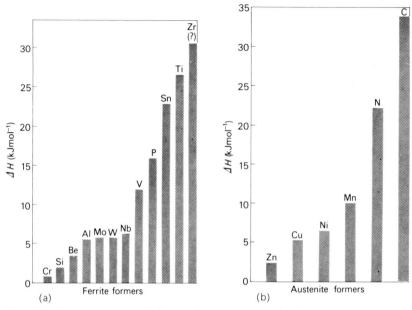

Fig. 4.3 Relative strength of alloying elements as: a, ferrite formers; b, austenite formers (Andrews, *JISI*, 1956, **184**, 414)

histograms illustrate the relative strengths of alloying elements in terms of ΔH. The ferrite formers are listed in (a) and the austenitic formers in (b).

4.2 The distribution of alloying elements in steels

Although only binary systems have been considered so far, when carbon is included to make ternary systems the same general principles usually apply. For a fixed carbon content, as the alloying element is added the γ-field is either expanded or contracted depending on the particular solute. With an element such as silicon the γ-field is restricted and there is a corresponding enlargement of the α-field. If vanadium is added, the γ-field is contracted and there will be vanadium carbide in equilibrium with ferrite over much of the ferrite field. Nickel does not form a carbide and expands the γ-field. Normally elements with opposing tendencies will cancel each other out at the appropriate combinations, but in some cases anomalies occur. For example, chromium added to nickel in a steel in concentrations around 18% helps to stabilize the γ-phase, as shown by 18Cr 8% Ni austenitic steels (Chapter 11).

One conveient way of illustrating quantitatively the effect of an alloying element on the γ-phase field of the Fe-C system is to project on to the Fe-C plane of the ternary system the γ-phase field boundaries for increasing concentration of a particular alloying element. This is illustrated in Fig. 4.4

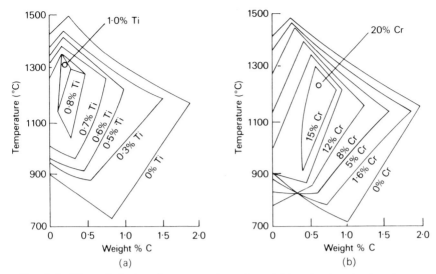

Fig. 4.4 Effect of alloying elements on the γ-phase field: a, titanium; b, chromium (after Tofaute and Büttinghaus, *Archiv. Eisenhüttenwesen*, 1938, **12**, 33)

for titanium and chromium, from which it can be seen that just over 1% Ti will eliminate the γ-loop, while 20% Cr is required to reach this point. Other ternary systems can be followed in the same way, for example in Fe-V-C, vanadium has an effect intermediate between that of titanium and of chromium.

For more precise and extensive information, it is necessary to consider series of isothermal sections in true ternary systems Fe-C-X, but even in some of the more familiar systems the full information is not available, partly because the acquisition of accurate data can be a difficult and very time-consuming process. Recently the introduction of computer-based methods has permitted the synthesis of extensive thermochemical and phase equilibria data, and its presentation in the form, for example, of isothermal sections over a wide range of temperatures. A journal† now publishes the work of laboratories concerned with such work, and of particular relevance is the recent detailed data on the Fe-Mn-C and Fe-Cr-C systems††.

If only steels in which the austenite transforms to ferrite and carbide on slow cooling are considered, the alloying elements can be divided into three categories:
(1) elements which enter only the ferrite phase
(2) elements which form stable carbides and also enter the ferrite phase

† *Calphad, Computer Coupling of Phase Diagrams and Thermochemistry*, Pergamon Press, Oxford
†† Hillert, M. and Waldenström, M., *Calphad*, 1977, **1**, 97 (Fe-Mn-C); Lundberg, R., Waldenström, M. and Uhrenius, B., *Calphad*, 1977, **1**, 159 (Fe-Cr-C)

60 *Steels—Microstructure and Properties*

(3) elements which enter only the carbide phase.

In the first category there are elements such as nickel, copper, phosphorus and silicon which, in transformable steels, are normally found in solid solution in the ferrite phase, their solubility in cementite or in alloy carbides being quite low.

The majority of alloying elements used in steels fall into the second category, in so far as they are carbide formers and as such, at low concentrations, go into solid solution in cementite, but will also form solid solutions in ferrite. At higher concentrations most will form alloy carbides, which are thermodynamically more stable than cementite. Typical examples are manganese, chromium, molybdenum, vanadium, titanium, tungsten and niobium. The stability of the alloy carbides and nitrides frequently found in steels relative to that of cementite is shown in Fig. 4.5, where the enthalpies of formation, ΔH, are plotted. Manganese carbide is

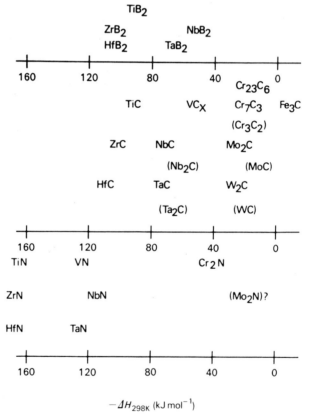

Fig. 4.5 Enthalpies of formation of carbides, nitrides and borides (after Schick, *Thermodynamics of Certain Refractory Compounds*, Academic Press, New York, 1966)

not found in steels, but instead manganese enters readily into solid solution in Fe_3C. The carbide-forming elements are usually present greatly in excess of the amounts needed in the carbide phase, which are determined primarily by the carbon content of the steel. The remainder enter into solid solution in the ferrite with the non-carbide forming elements nickel and silicon. Some of these elements, notably titanium, tungsten, and molybdenum, produce substantial solid solution hardening of ferrite.

In the third category there are a few elements which enter predominantly the carbide phase. Nitrogen is the most important element and it forms carbo-nitrides with iron and many alloying elements. However, in the presence of certain very strong nitride forming elements, e.g. titanium and aluminium, separate alloy nitride phases can occur.

While ternary phase diagrams, Fe-C-X, can be particularly helpful in understanding the phases which can exist in simple steels, isothermal sections for a number of temperatures are needed before an adequate picture of the equilibrium phases can be built up. For more complex steels the task is formidable and equilibrium diagrams can only give a rough guide to the structures likely to be encountered. It is, however, possible to construct pseudobinary diagrams for groups of steels, which give an overall view of the equilibrium phases likely to be encountered at a particular temperature. For example, Cr-V steels are widely used in the heat-treated

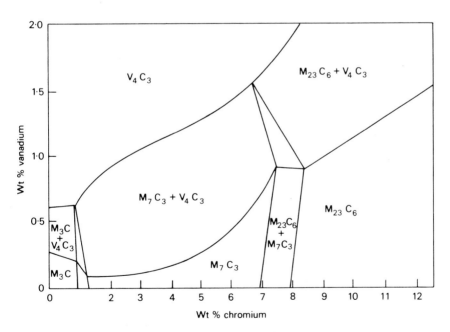

Fig. 4.6 Carbide constitution in 0.2%C steels at 700°C as a function of vanadium and chromium content (Shaw and Quarrell, *JISI*, 1957, **185**, 10)

condition, and both chromium and vanadium are carbide formers. If a particular carbon level, e.g. 0.2 wt% and a temperature at which equilibrium can be readily reached, e.g. 700° C, is chosen, it is possible to examine a wide range of different compositions to identify the carbide phases in equilibrium with ferrite at that temperature. The phase fields can then be plotted on a diagram as a function of chromium and vanadium, as shown in Fig. 4.6. It should be noted that cementite is only stable up to about 1.5% chromium or 0.6% vanadium and, for much of the diagram, several alloy carbides replace cementite.

4.3 The effect of alloying elements on the kinetics of the γ/α transformation

Since alloying elements have different tendencies to exist in the ferrite and carbide phases, it might be expected that the rate at which the decomposition of austenite occurs below Ae_1 would be sensitive to the concentration of alloying elements in a steel. Both the growth of ferrite and pearlite are affected, so these reactions will be considered separately. Most familiar alloying elements displace the *TTT* curve for a plain carbon steel to the right, i.e. towards longer transformation times. However, a small group of elements move the curve to shorter transformation times.

4.3.1 The effect of alloying elements on the ferrite reaction

Two basically different modes of growth of pre-eutectoid ferrite in austenite have been observed in Fe-C-X alloys. The actual mode observed is dependent on the composition of the alloy but the two modes may occur at different temperatures in the same alloy. The modes are
(1) growth with partition of the alloying element X between α and γ under local equilibrium conditions
(2) growth with no partition of X between α and γ.
In the first mode, the ferrite is growing at a slow rate determined by the diffusivity of the alloying element X in the γ-phase. This behaviour is sensitive to alloy composition which is shown by the fact that an Fe-1.3 at% C-3.2 at% Mn alloy exhibits Mn partition at 742° C, whereas an Fe-1.0 at% C-1.5 at% Mn alloy shows partition of manganese at 725° C. Alloys in which X is Ni or Pt also show partition at higher transformation temperatures.

The mode where no partition occurs is much more common. This mode involves a narrow zone of enrichment or depletion, depending on whether X is a γ- or α-stabilizer, which moves ahead of the α/γ interface. Aaronson and Domian have shown this to occur for alloys in which X = Si, Mo, Co, Al, Cr and Cu for all temperatures investigated. It is notable that the strong γ-stabilizers Ni, Mn and Pt are the only elements to show partition to the γ-phase. In the no-partition regime the observed growth rates are relatively high, being determined by the diffusivity of carbon which diffuses several orders of magnitude faster than the metallic alloying elements. However, it has been shown that in Fe-C-X alloys, ferrite still grows much more slowly than in Fe-C alloys, even when no partition of X is observed. This is illustrated in Fig. 4.7 from the work of Kinsman and Aaronson for X = Mn and Mo. To explain these results they proposed that the α/γ boundary collects X atoms during the transformation and, as a result, experiences an *impurity drag*.

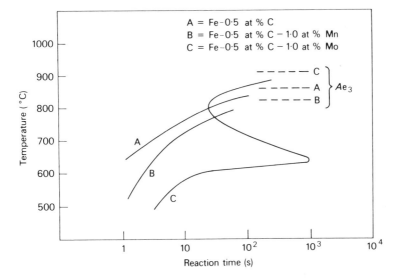

Fig. 4.7 Effect of manganese and molybdenum on the kinetics of the ferrite reaction (Kinsman and Aaronson, In: *Transformation and Hardenability in Steels*, Climax Molybdenum Co., 1967)

A third approach to the ferrite reaction was introduced by Hultgren, who proposed a state of para-equilibrium at the γ/α boundary, i.e. the α and γ are in local equilibrium with the fast moving species carbon but the alloying element X is inert, holding its ratio constant across the interface. This is a metastable mode which allows growth of ferrite controlled by the diffusivity of carbon, with no partition at all of alloying element X.

4.3.2 The effect of alloying elements on the pearlite reaction

The pearlite reaction is a typical nucleation and growth reaction and, under the appropriate experimental conditions, rates of nucleation $\dot{N}$ and rates of growth G can be determined (see Chapter 3). The work of Mehl and coworkers showed that many alloying elements reduce both $\dot{N}$ and G. For example, in molybdenum steels of eutectoid composition both $\dot{N}$ and G

were decreased, and nickel steels behaved in a similar manner. The growth rate G as a function of atomic concentration of alloying elements in several groups of steels is shown in Fig. 4.8. The change in slope for Mo steels was correlated with the substitution of cementite by a molybdenum-rich carbide. Certain elements, notably cobalt, increased both $\dot{N}$ and G for the pearlite reaction. The rates of growth of pearlite nodules at 660° C in cobalt steels are compared with that of a Co-free steel in Fig. 4.9.

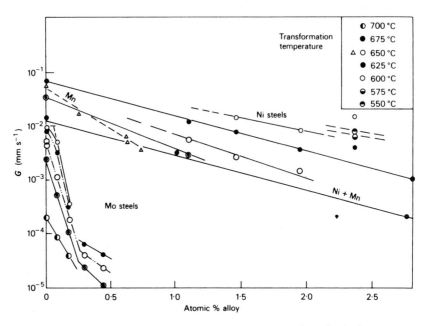

Fig. 4.8 Effect of alloying elements on the rate of growth of pearlite in the range 550–700°C (Mehl and Hagel, *Progress in Metal Physics*, 1956, **6**, 74)

Recent work on chromium steels has shown that the addition of 1% Cr to a eutectoid steels results in substantially lower growth rates of pearlite. It follows that in general the C-curve for a pearlitic steel will be moved to longer times as the concentration of alloying element is increased.

On examining the interface between pearlite and austenite during transformation, it appears that the basic nature of the pearlite reaction requires partition of carbon between the cementite and ferrite (Chapter 3). However, in the presence of metallic alloying elements, it is not obvious, *ab initio*, whether partition of these elements will take place, taking partition to mean partition at the pearlite/austenite interface so that element X partitions between cementite and ferrite as they are formed. At a later stage of the reaction, and after its completion, alloying elements can partition within the pearlite over a wide temperature range.

It is now generally agreed that partition of X between cementite and

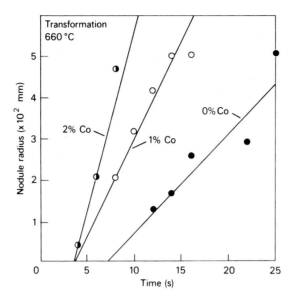

Fig. 4.9 Effect of cobalt on pearlite growth rate (Mehl and Hagel, *Progress in Metal Physics*, 1956, **6**, 74)

ferrite at the interface with austenite does occur in some systems, provided the transformation temperature is high enough. While partition can be predicted on theoretical grounds, it can now also be investigated experimentally† using electron probe microanalysis, where a probe size of < 0.1 μm allows the *in-situ* analysis of pearlitic ferrite and cementite in partly transformed alloys. In this way the systems Fe-Mn-C and Fe-Cr-C have been examined, and partition found in the former above 683° C for a 1.08 Mn alloy and 650° C for a 1.80 Mn alloy, and in the latter above 700° C.

An approach to the pearlite reaction, similar to that described earlier for the ferrite reaction, is to distinguish two modes of growth, a partition local equilibrium and a non-partition local equilibrium situation††, which are both temperature and composition dependent. Elements which favour the formation of austenite, and so depress the eutectoid temperature, and also have low solubilities in cementite, e.g. Ni, will encourage the non-partitioning reaction. Those elements which are strong ferrite formers and consequently raise the eutectoid temperature, as well as being soluble in cementite, are likely to exhibit the partitioning type of reaction at the higher transformation temperatures, e.g. Cr, Mo, Si. The growth of pearlite in the non-partitioning case is probably controlled by volume diffusion of carbon in austenite, but this diffusivity is reduced by the presence of other alloying

† Razik, N. A., Lorimer, G. W. and Ridley, N., *Acta Met.*, 1974, **22**, 1249
†† Coates, D. E., *Met. Trans.*, 1973, **4**, 2313

elements, accounting for the observed effect of elements such as Ni on the pearlite growth rate. Where partitioning of X takes place, the diffusivity of the alloying element in austenite must be a limiting factor.

Whatever the alloying element distribution is at the growing interface, subsequent redistribution between the ferrite and cementite takes place, i.e. those elements with substantial solubility in cementite (carbide formers) will diffuse into that phase and the non-carbide formers will not. In this way the composition of cementite can vary over wide limits, e.g. manganese is very soluble in Fe_3C; up to 20% of the iron atoms can be replaced by chromium, while vanadium will replace 10% and molybdenum only 4%. The change in composition of cementite, while not affecting the crystal structure, will influence, for example, the pearlite interlamellar spacing, the detailed morphology and the tendency to spheroidize.

Once the alloying element concentration reaches a critical level, the cementite will be replaced by another carbide phase. For example, in a chromium, tungsten or molybdenum steel, the complex cubic $M_{23}C_6$ carbide can form, where M can include iron, chromium, molybdenum or tungsten (Figs. 4.10 and 4.11). This change in the carbide phase does not necessarily alter the basic pearlitic morphology and consequently alloy pearlites are obtained in which an alloy carbide is associated with ferrite

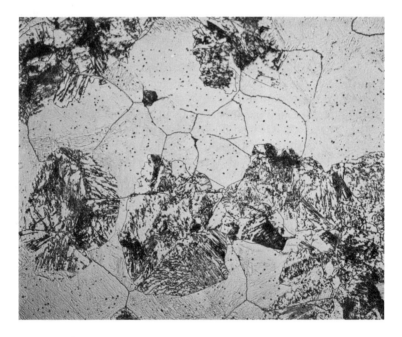

Fig. 4.10 Fe-12Cr-0.2C transformed 30 min at 775°C. Pearlite-type reaction involving $M_{23}C_6$ (Campbell). Optical micrograph ×300

(Fig. 4.11). These pearlites occur only in medium and highly-alloyed steels, usually at the highest transformation temperatures. At lower transformation temperatures in the same steel, cementitic pearlite may still form because of the inadequate diffusion of the alloying element.

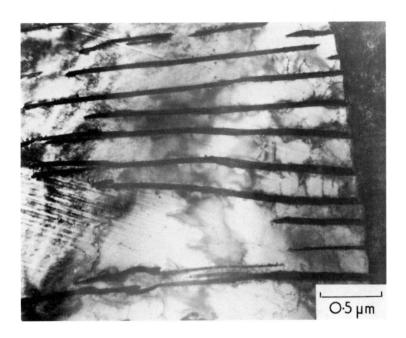

Fig. 4.11 Fe-12Cr-0.2C transformed 15 min at 750°C. Alloy pearlite, $M_{23}C_6$/ferrite (Campbell). Thin-foil EM

4.4 Structural changes resulting from alloying additions

The addition to iron-carbon alloys of elements such as nickel, silicon, manganese, which do not form carbides in competition with cementite, does not basically alter the microstructures formed after transformation. However, in the case of strong carbide-forming elements such as molybdenum, chromium and tungsten, cementite will be replaced by the appropriate alloy carbides, often at relatively low alloying element concentrations. Still stronger carbide forming elements such as niobium, titanium and vanadium are capable of forming alloy carbides, preferentially at alloying concentrations less than 0.1 wt%. It would, therefore, be expected that the microstructures of steels containing these elements would be radically altered.

It has been shown how the difference in solubility of carbon in austenite and ferrite leads to the familiar ferrite/cementite aggregates in plain carbon steels. This means that, because the solubility of cementite in austenite is much greater than in ferrite, it is possible to redistribute the cementite by holding the steel in the austenite region to take it into solution, and then allowing transformation to take place to ferrite and cementite. Examining the possible alloy carbides, and nitrides, in the same way, shows that all the familiar ones are much less soluble in austenite than is cementite. In Fig. 4.12 the solubility products in austenite of vanadium, titanium and

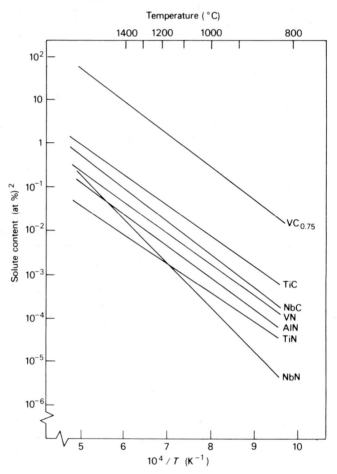

Fig. 4.12 Solubility products of carbides and nitrides in austenite as a function of temperature (Aronsson, In: *Steel Strengthening Mechanisms*, Climax Molybdenum Co., 1969)

niobium carbides and nitrides are plotted as a function of $\frac{1}{T}$. Chromium and molybdenum carbides are not included, but they are substantially more soluble in austenite than the other carbides. Detailed consideration of such data, together with practical knowledge of alloy steel behaviour, indicates that, for niobium and titanium, concentrations of greater than about 0.25 wt % will form excess alloy carbides which cannot be dissolved in austenite at the highest solution temperatures. With vanadium the limit is higher at 1–2%, and with molybdenum up to about 5%. Chromium has a much higher limit before complete solution of chromium carbide in austenite becomes difficult. This argument assumes that sufficient carbon is present in the steel to combine with the alloying element. If not, the excess metallic element will go into solid solution both in the austenite and the ferrite.

4.4.1 Ferrite/alloy carbide aggregates

Steels containing strong carbide-forming elements transform from austenite to ferrite in a similar way to, for example, steels containing nickel or silicon. However, the carbide-forming elements restrict very substantially the γ-loop (Fig. 4.4), so that the eutectoid composition is depressed to much lower carbon levels and to higher transformation temperatures. One result is that pearlite can completely disappear from the transformed microstructures, which now exhibit very different ferrite/carbide aggregates, usually on a very much finer scale than pearlite. Apart from the alloy carbide-pearlites, particularly found in high chromium steels, there are three morphologies of alloy carbides which are intimately associated with ferrite in the transformation temperature range in which plain carbon steels form ferrite/pearlite structures.

Continuous growth of fibres/laths The alloy carbides form as fine fibres or laths which grow normal to the γ/α interface, and continue to grow as the interface moves forward forming fibrous aggregates of carbide and ferrite (Fig. 4.13).

Repeated nucleation of carbides (interphase precipitation) In this growth mode the carbide particles, usually in the form of small plates or rods, nucleate at the γ/α interface which then moves to a new position where the nucleation cycle again occurs. This process can be repeated many hundreds of times within a particular austenite grain leading to a ferrite matrix with very fine banded dispersions as, for example, in the 0.75% vanadium steel shown in Fig. 4.14. Chromium steels give coarser dispersions (Fig. 4.16).

Nucleation in supersaturated ferrite In microalloyed steels, where strong carbide-forming elements are present in concentrations less than 0.1 wt%, it is often possible to obtain the ferrite in a supersaturated condition with little or no carbide precipitation taking place during the γ/α transformation. Instead, while the steel is held at the transformation temperature, carbide precipitates form within the newly-formed ferrite grains, usually on dislocations (Fig. 4.15).

70 Steels—Microstructure and Properties

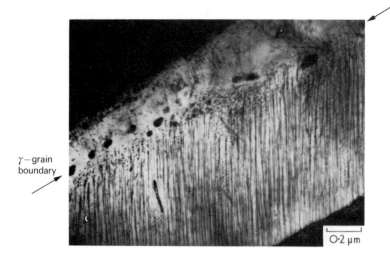

Fig. 4.13 Fe-4Mo-0.2C transformed 20 min at 650°C. Fibrous Mo_2C growth from γ boundary (Berry). Thin-foil EM

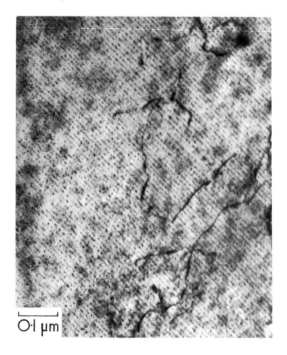

Fig. 4.14 Fe-0.75V-0.15C transformed 5 min at 725°C. Interphase precipitation of VC in ferrite (Batte). Thin-foil EM

The effects of alloying elements on iron-carbon alloys 71

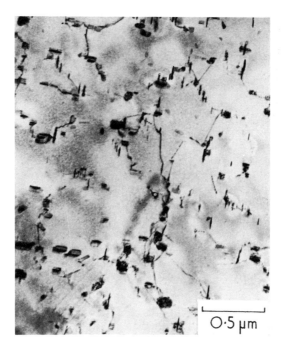

Fig. 4.15 Fe-0.25V-0.05C transformed and held $2\frac{1}{2}$ h at 740°C. VC precipitation on dislocations (Balliger). Thin-foil EM

While it is possible by careful choice of alloy and experimental conditions to obtain each of the above microstructures separately, in practice they are often all present in transformed alloy steels, provided the steel contains a strong carbide-forming element. Consequently the microstructures of transformable alloy steels can be very complex, the full extent of these complexities only being revealed when high resolution electron microscopy is used to study them.

In general, the fibrous morphology represents a closer approach to an equilibrium structure so it is more predominant in steels which have transformed slowly. In contrast, the interphase precipitation and dislocation nucleated structures occur more readily in rapidly transforming steels, where there is a high driving force, for example, in microalloyed steels (Chapter 9).

4.4.2 Alloy carbide fibres and laths

The clearest analogy with pearlite is found when the alloy carbide in lath morphology forms nodules in association with ferrite. These pearlitic nodules are often encountered at temperatures just below Ae_1 in steels which transform relatively slowly. For example, these structures are obtained in chromium steels with between 4% and 12% chromium (Fig. 4.11), and the crystallography is analogous to that of cementitic pearlite. It is, however, different in detail because of the different crystal structures of the possible carbides, e.g. Cr_7C_3 is hexagonal and $Cr_{23}C_6$ is complex cubic. The structures observed are relatively coarse, but finer than pearlite formed under equivalent conditions, because of the need for the partition of the alloying element, e.g. chromium between the carbide and the ferrite. To achieve this, the interlamellar spacing must be substantially finer than in the equivalent iron-carbon case.

At lower temperatures the lath morphology is largely replaced by much finer fibrous aggregates, e.g. in high Cr steels coarse laths of $Cr_{23}C_6$ can be replaced by fine fibres of the same carbide usually 500 Å in diameter. In steels with 4% molybdenum, transformed in the range 850–600° C, fibrous carbide-ferrite aggregates are the predominant morphology in which the Mo_2C fibres vary between 100 and 300 Å diameter. Their length, which is determined by the size of the ferrite colony, can be up to 10 μm with little or no branching. Similar morphologies occur, but are much less dominant, in steels containing W, Ti, V and Nb.

Carbide fibres are frequently associated with planar interfaces, as well as with pearlitic-type interfaces. Nevertheless, these are boundaries which can apparently propagate rapidly without the need for step migration. A recent detailed computer analysis of similar boundaries in austenitic steels has shown that they possess comparatively high densities of coincident lattice sites[†].

4.4.3 Interphase precipitation

Interphase precipitation has been shown to nucleate periodically at the γ/α interface during the transformation. The precipitate particles form in bands which are closely parallel to the interface, and which follow the general direction of the interface even when it changes direction sharply. A further characteristic is the frequent development of only one of the possible Widmanstätten variants, for example VC platelets in a particular region are all only of one variant of the habit, i.e. that in which the plates are most nearly parallel to the interface. The bands are often associated with planar low energy interfaces, and the interband spacing is determined by the height of steps which move along the interface (Fig. 4.16). The nucleation of the carbide particles occurs normally on the low energy planar interfaces, rather than on the rapidly moving high energy steps.

The need for step movement on the γ/α interface is in contrast to the growth of fibrous carbides behind an interface on which no steps are

---

† Ainsley, M. H., Cocks, G. J. and Miller, D. R., *Metal Science*, 1979, **13**, 20

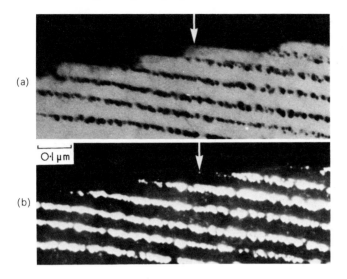

Fig. 4.16 Fe-12Cr-0.2C transformed 30 min at 650°C. Precipitation of $M_{23}C_6$ at stepped γ/α interface: a, bright field; b, precipitate spot dark field (Campbell). Thin-foil EM

observed. Indeed if, in these circumstances, a step does move along the interface, the fibrous growth is stopped and replaced by interphase precipitation. The step height and, therefore, the band spacing of the precipitation, is dependent on the temperature of transformation and on the composition. As the temperature of transformation is lowered the band spacing is reduced, e.g. in a 1% V 0.2% carbon steel, the spacing varies from 25 nm at 825°C to 7.5 nm at 725°C (Fig. 4.17), and at lower temperatures has been observed to be less than 5 nm. The extremely fine scale of this phenomenon in vanadium steels, which also occurs in Ti and Nb steels, is due to the rapid rate at which the γ/α transformation takes place. At the higher transformation temperatures, the slower rate of reaction leads to coarser structures. Similarly, if the reaction is slowed down by addition of further alloying elements, e.g. Ni and Mn, the precipitate dispersion coarsens. The scale of the dispersion also varies from steel to steel, being coarsest in chromium, tungsten and molybdenum steels where the reaction is relatively slow, and much finer in steels in which vanadium, niobium and titanium are the dominant alloying elements and the transformation is rapid.

4.4.4 Nucleation in supersaturated ferrite

It has been shown that ferrite can occur in different morphologies depending on the transformation temperature. At the highest transfor-

74 Steels—Microstructure and Properties

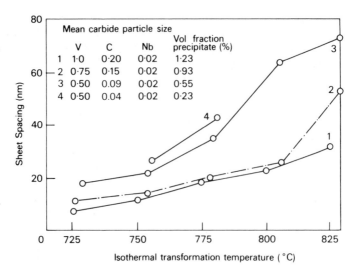

Fig. 4.17 Interphase precipitation of VC in vanadium steels. Precipitate sheet spacing as a function of transformation temperature

mation temperatures, equi-axed boundary allotriomorphs form at the austenite grain boundaries, and carbon diffuses to the austenite. In alloy steels, e.g. low V steels, there is evidence that the alloying element can also partition. As a result no alloy carbide forms in this ferrite, which is thus truly pro-eutectoid. At lower temperatures the ferrite formed is still equi-axed, but the alloy carbide forms at the same time either as interphase precipitate or as fibres. This is probably the closest approach to true eutectoid behaviour in an alloy steel containing a strong carbide-forming element.

At still lower transformation temperatures the ferrite adopts a Widmanstätten habit and forms as laths, as in pure iron-carbon alloys. However, this ferrite can be supersaturated when first formed. If held only for a short time at the transformation temperature, precipitation of the alloy carbide occurs within the ferrite on dislocations. Such behaviour would be expected in alloy steels with acicular ferrite provided a strong carbide former such as V, Ti or Nb is present although, in theory, similar structures should be possible in plain carbon steels.

4.5 Transformation diagrams for alloy steels

The transformation of austenite below the eutectoid temperature can best be presented in an isothermal transformation diagram (Chapter 3), in which the beginning and end of transformation is plotted as a function of temperature and time. Such curves are known as time-temperature-transformation, or *TTT*, curves and form one of the important sources of quantitative information for the heat treatment of steels. In the simple case of a eutectoid plain carbon steel, the curve is roughly C-shaped with the pearlite reaction occurring down to the nose of the curve and a little beyond. At lower temperatures bainite and martensite form (see Chapters 5 and 6). The diagrams become more complex for hypo- and hyper-eutectoid alloys as the ferrite or cementite reactions have also to be represented by additional lines.

Alloying elements, on the whole, retard both the pro-eutectoid reactions and the pearlite reaction, so that *TTT* curves for alloy steels are moved increasingly to longer times as the alloy content is increased. Additionally, those elements which expand the γ-field depress the eutectoid temperature, with the result that they also depress the position of the *TTT* curves relative to the temperature axes. This behaviour is shown by steels containing manganese or nickel. For example, in a 1.3 % Mn 0.8 % C steel, pearlite can form at temperatures as low as 400°C. In contrast, elements which favour the ferrite phase raise the eutectoid temperature and the *TTT* curves move correspondingly to higher temperatures. The slowing down of the ferrite and pearlite reactions by alloying elements enables these reactions to be more readily avoided during heat treatment, so that the much stronger low temperature phases such as bainite and martensite can be obtained in the microstructure. The hard martensitic structure is only obtained in plain carbon steels by water quenching from the austenitic condition whereas, by the addition of alloying elements, a lower critical cooling rate is needed to achieve this condition. Consequently, alloy steels allow hardening to occur during oil quenching, or even on air cooling, if the *TTT* curve has been sufficiently displaced to longer times.

Further reading

Hume-Rothery, W., *The Structure of Alloys of Iron*, Pergamon Press, Oxford, 1966

Andrews, K. W., *Metal Treatment*, **19**, 425; 489, 1952; *Iron and Steel*, March 1961

Goldschmidt, H. J., *Interstitial Alloys*, Butterworths, 1967

Bain, E. C. and Paxton, H. W., *Alloying Elements in Steel*, American Society for Metals, 1961

Bullens, D. K., *Steel and Its Heat Treatment*, Volumes 1 and 2, Wiley, 1956

Kumar, R., *Physical Metallurgy of Iron and Steel*, Asia Publishing House, 1968

AIME, Symposium on cellular and pearlite reactions, Detroit 1971. *Met. Trans.*, **3**, 2717–2804, 1972

Honeycomb, R. W. K., Ferrite, Hatfield Memorial Lecture 1979, *Metal Science*, **14**, 1980

5
The formation of martensite

5.1 Introduction

Rapid quenching of austenite to room temperature often results in the formation of martensite, a very hard structure in which the carbon, formerly in solid solution in the austenite, remains in solution in the new phase. Unlike ferrite or pearlite, martensite forms by a sudden shear process in the austenite lattice which is not normally accompanied by atomic diffusion. Ideally, the martensite reaction is a diffusionless shear transformation, highly crystallographic in character, which leads to a characteristic lath or lenticular microstructure. The martensite reaction in steels is the best known of a large group of transformations in alloys in which the transformation occurs by shear without change in chemical composition. The generic name of martensitic transformation describes all such reactions. It should however be mentioned that there is a large number of transformations which possess the geometric and crystallographic features of martensitic transformations, but which also involve diffusion. Consequently, the broader term of shear transformation is perhaps best used to describe the whole range of possible transformations.

5.2 General characteristics

The martensite reaction in steels normally occurs athermally, i.e. during cooling in a temperature range which can be precisely defined for a particular steel. The reaction begins at a *martensitic start temperature*, M_s, which can vary over a wide temperature range from as high as 500°C to well below room temperature, depending on the concentration of γ-stabilizing alloying elements in the steel. Once the M_s is reached further transformation takes place during cooling until the reaction ceases at the M_f temperature. At this temperature all the austenite should have transformed to martensite but frequently, in practice, a small proportion of the austenite does not transform. This is called *retained austenite*. Larger volume fractions of austenite are retained in some highly alloyed steels, where the M_f temperature is well below room temperature. To obtain the martensitic reaction it is usually necessary for the steel to be rapidly cooled, so that the metastable austenite reaches M_s. The rate of cooling must be sufficient to suppress the higher temperature diffusion-controlled ferrite and pearlite reactions, as well as other intermediate reactions such as the formation of bainite. The critical rate of cooling required is very sensitive to the alloying elements present in the steel and, in general, will be lower the higher total alloy concentration.

The martensite reaction is readily recognized by its appearance in the optical microscope. Each grain of austenite transforms by the sudden formation of thin plates or laths of martensite of striking crystallographic character (Fig. 5.1). The laths have a well-defined habit plane (Fig. 5.2) and they normally occur on several variants of this plane within each grain. The habit plane is not constant, but changes as the carbon content is increased.

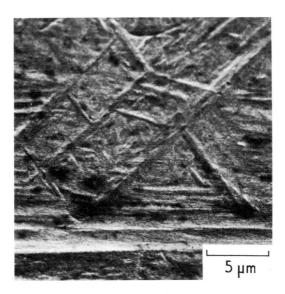

Fig. 5.1 Martensite in a 4Ni, 0.4C steel shown in relief on a pre-polished surface (Bhadeshia). Stereoscan, ×3400

For example, in lower carbon steels the habit phase is usually $\{111\}_\gamma$, while for 0.5 to 1.4% carbon, the usual plane observed is $\{225\}_\gamma$ and at very high carbon levels (above 1.4%) the plane changes to $\{259\}_\gamma$. There is also an orientation relationship between the new martensitic lattice which has a tetragonal structure and the austenite. For lower carbon steels, the orientation relationship is that due to Kurdjumov and Sachs, already shown to occur between austenite and ferrite, i.e.

$\{111\}_\gamma // \{110\}_{\alpha'}$, $\{111\}_\gamma$ habit plane
$\langle 011 \rangle // \langle 111 \rangle_{\alpha'}$ (α' = martensite)

This orientation relationship persists when the habit plane changes to

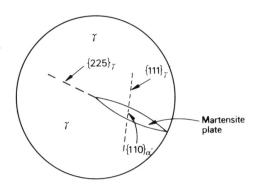

Fig. 5.2 Schematic diagram of typical crystallography of a martensite plate

{225}. However, when the {259} habit plane appears, a new relationship attributed to Greninger and Troiano and to Nishiyama is found, i.e.

$$\{111\}_\gamma // \{110\}_{\alpha'}$$
$$\langle 112 \rangle_\gamma // \langle 011 \rangle_{\alpha'} \quad \{259\}_\gamma \text{ habit plane}$$

It should be pointed out that neither of these relationships is precise, and habit planes show a scatter of several degrees about the ideal orientation.

Evidence that the martensitic transformation involves lattice shears can easily be obtained by polishing a surface, or better two surfaces at right angles, prior to transformation. After martensite plates form, the surface reveals shear displacements easily observed by the local relief that occurs (Fig. 5.1), and by displacements shown by surface scratches (Fig. 5.3). These experiments show that a martensitic reaction is accompanied by a *shape change*. Such changes can be analysed by use of two beam interferometry, and quantitative data obtained from the displacement of the fringe patterns.

5.3 The crystal structure of martensite

Martensite is a supersaturated solid solution of carbon in iron which has a body-centred tetragonal (bct) structure, a distorted form of bcc iron. The tetragonality measured by the ratio between the axes, $\frac{c}{a}$, increases with carbon content:

$$\frac{c}{a} = 1 + 0.045 \text{ wt\%C} \qquad (5.1)$$

implying that at zero carbon content the structure would be bcc, free of distortion. The effect of carbon on the lattice parameter of austenite, and on the c and a parameters of martensite is shown in Fig. 5.4.

It is interesting to note that carbon in interstitial solid solution expands

Fig. 5.3 Fe-30.5Ni-0.3C, illustrating the displacement of surface scratches by the martensitic shear (Bhadeshia). Nomanski interference contrast, ×650

the fcc iron lattice uniformly, but with bcc iron the expansion is nonsymmetrical giving rise to tetragonal distortion. To understand this important difference in behaviour, it is necessary to compare the interstitial sites for carbon in the two lattices. In each case, carbon atoms occupy octahedral sites, indicated for martensite in black in Fig. 5.5, and have six near-neighbour iron atoms. In the fcc lattice the six iron atoms around each interstitial carbon atoms form a regular octahedron, whereas in the bcc case the corresponding octahedra are not regular, being shortened along the z-axis (Fig. 1.2e,f). These compressed octahedra only have four-fold symmetry along the shortened axis in each case, in contrast to the fcc structure in which the regular octahedra have three four-fold axes of symmetry.

Analysis of the distortion produced by carbon atoms in the several types of site available in the fcc and bcc lattices, has shown that in the fcc structure the distortion is completely symmetrical, whereas in the bcc one, interstitial atoms in z positions will give rise to much greater expansion of iron-iron atom distances than in the x and y positions. Assuming that the fcc → bcc tetragonal transformation occurs in a diffusionless way, there will be no opportunity for carbon atoms to move, so those interstitial sites already occupied by carbon will be favoured. Since only the z sites are common to both the fcc and bcc lattices, on transformation there are more carbon atoms at these sites causing the z-axis to expand, and the non-regular

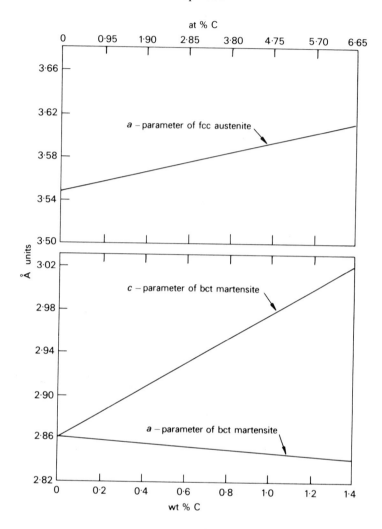

Fig. 5.4 Effect of carbon on the lattice parameters of austenite and of martensite (after Roberts, In: Cohen, *Trans. Met. Soc. AIME*, 1962, **224**, 638)

octahedron becomes more regular. This is largely a unidirectional distortion which leads to the bc tetragonal lattice, the z axis now corresponding to the c-axis in the tetragonal lattice.

Therefore, the tetragonality of martensite arises as a direct result of interstitial solution of carbon atoms in the bcc lattice, together with the preference for a particular type of octahedral site imposed by the diffusionless character of the reaction.

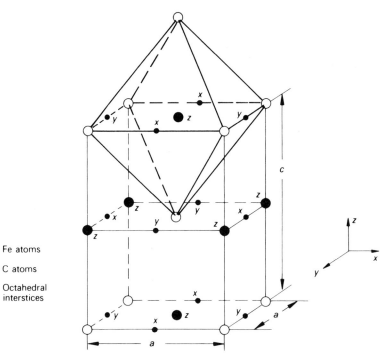

Fig. 5.5 Martensite body-centred tetragonal lattice illustrating the three sets of octahedral interstices. The z-set is fully occupied by carbon atoms (Cohen, *Trans. Met. Soc. AIME*, 1962, **224**, 638)

5.4 The crystallography of martensitic transformations

So far we have not considered the orientation change which takes place when a martensite plate is formed inside an austenite grain. While the observed orientation relationships can change with composition, in all ferrous systems the closest packed planes are approximately parallel, i.e. $\{111\}_\gamma$ is parallel to $\{110\}_{\alpha'}$. The two main relationships are then described by the parallelism of different lattice directions, amounting to several degrees difference in orientation.

$\langle 011 \rangle_\gamma // \langle 111 \rangle_{\alpha'}$ Kurdjumov-Sachs
$\langle 112 \rangle_\gamma // \langle 011 \rangle_{\alpha'}$ Nishiyama/Greninger & Troiano

The correspondence between the lattices of austenite and martensite was first pointed out by Bain, who showed that a tetragonal unit cell could be outlined within two unit cells of austenite (Fig. 5.6). To convert this cell into a martensite cell, a deformation (Bain strain) is necessary which involves a contraction of about 17% along the $[001]_\gamma$ corresponding to the c-axis of the martensite cell, and a uniform expansion of about 12% in the $(001)_\gamma$. This correspondence does not imply a mechanism for the phase

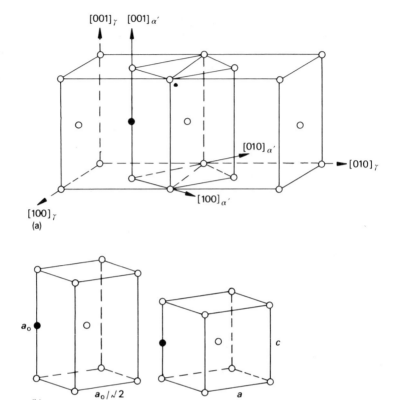

Fig. 5.6 The lattice correspondence for formation of martensite from austenite: a, tetragonal unit cell outlined in austenite; b, lattice deformation (compression along c-axis) to form martensite of correct c/a ratio (Bain strain) (Christian, In: *Martensite: Fundamentals and Technology*, ed. E. R. Petty, Longmans, 1970)

transformation, nor does it predict a habit phase or orientation relationship. Indeed, the observations of surface relief as a result of the transformation indicate that the principal strains required by the Bain distortion are inconsistent with experiments, which have revealed that the habit plane between the austenite and martensite remains undistorted during the transformation.

The observation of scratches on polished surfaces after transformation indicates that the martensite plates are apparently tilted about their junction plane with the austenite, leading to a *shape deformation* in which the junction plane of the martensite and austenite, i.e. the habit plane, undergoes no rotation. The fact that the scratches simply change direction at the interface implies that a homogeneous deformation or shear has taken place, and the continuity of scratches at the interface indicates that the habit plane has undergone no large-scale distortion. The shape change which allows the interface to remain planar and invariant during deformation is referred to as an *invariant plane strain*. In higher carbon ferrous martensites, the interface does not remain planar due to accommodation stresses in the matrix, with the result that the martensitic plates are lenticular in form (Fig. 5.7).

Fig. 5.7 Fe-1.8C-3Mn-2Si. Lenticular martensite illustrating the burst phenomenon (Bhadeshia). Optical micrograph, ×300

The existence of a lattice correspondence is the hallmark of a martensitic or shear transformation. In simple terms, this means that geometrically each atom in the parent phase can be related to a corresponding atom in the product phase, but the corresponding vectors in each lattice are not equal and angular relationships between planes are not preserved (Fig. 5.8). However, there are three mutually perpendicular vectors in the parent phase, the directions of which are unaffected by the transformation. These vectors are used to define the deformation required to create the new lattice, in so far as displacements which take place parallel to them are the *principal strains*. The correspondence applying in a particular transformation is usually the one resulting in the smallest atomic displacements and consequently the lowest strain energy. This will result in the smallest principal strains. This simple example shows that a correspondence is only exact for cases where the unit cells, i.e. heavy lines in Fig. 5.8, in parent and product have the same number of atoms. It is easy to visualize more

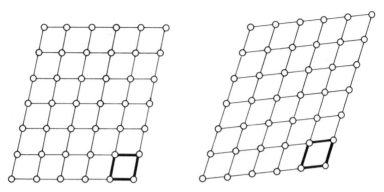

Fig. 5.8 A simple lattice correspondence (Christian, In: *Martensite: Fundamentals and Technology*, ed. E. R. Petty, Longmans, 1970)

complex correspondences where the lattice formed only defines the positions of a proportion of the atoms in the original lattice. Consequently there have to be local adjustments of atoms to conform to the new structure.

The homogeneous lattice deformation which generates the martensite lattice is referred to as a *pure strain*, and in ferrous alloys is the Bain strain previously referred to. However, in martensite, a basic difficulty which occurred early in the theoretical development was that the original matrix and the martensite structures did not match up along a lattice plane although a habit plane was always observed. Greninger and Troiano pointed out that a *shape deformation* resulting in invariant plane strain was insufficient to produce the observed result, and proposed that this homogeneous shape deformation occurring over large distances was complemented by an inhomogeneous *lattice deformation* occurring on an atomic scale. The lattice deformation was able to achieve accommodation at the interface, and this was assumed to occur without changing the nature of the product lattice, so it is strictly described as a *lattice invariant deformation*. This approach was further developed by Bowles and McKenzie, and by Wechsler, Lieberman and Read who independently developed formal *two-shear* theories.

For the ferrous martensites (the theories were applied to a wide range of martensites) the Bain strain was assumed to occur to achieve the new lattice of martensite from austenite, because it involves the smallest atomic displacements, and should therefore satisfy a minimum strain-energy criterion. This is the homogeneous shape deformation which, when combined with a fine scale inhomogeneous deformation (lattice invariant), led to the interface being undistorted. The inhomogeneous lattice invariant deformation was assumed to result from dislocation movement which could be in the form of deformation by slip or by twinning. Both these processes are equally capable of accommodating the misfit, which would otherwise occur at the austenite/martensite interface as a result of the deformation which causes both the macroscopic shape change, and the transformation of the original lattice from face-centred cubic to body-centred tetragonal. Fig. 5.9 shows schematically the two types of lattice invariant deformation occurring within a martensite plate. It should be noted that the block of martensite formed has produced a surface tilt and that the observed habit is preserved by the accommodation provided by either slip (Fig. 5.9a) or twinning (Fig. 5.9b). The result is a macroscopically planar interface which would clearly have irregularities on a very fine scale.

The above theoretical approach had considerable success in predicting the observed habit planes, the orientation relationships between matrix and

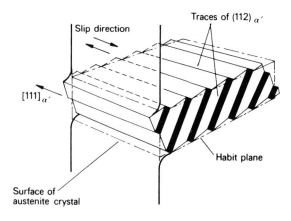

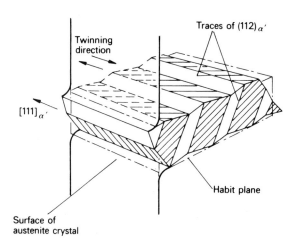

Fig. 5.9 Formation of martensite plate, illustrating two types of lattice deformation: a, slip; b, twinning (Christian, In: *Martensite: Fundamentals and Technology*, ed. E. R. Petty, Longmans, 1970)

the martensite, as well as the shape deformation for a number of martensitic transformations including ferrous martensites. It is, however, necessary to have accurate data, so that the habit planes of individual martensite plates can be directly associated with a specific orientation relationship of the plate with the adjacent matrix. For example, Greninger and Troiano used an iron-22% nickel 0.8% carbon alloy in which austenite was retained in association with martensite to predict successfully the correct habit plane, which in this alloy is an irrational plane near {3, 10, 15}, and also the shape change and the orientation relationship between martensite and austenite.

5.5 The morphology of ferrous martensites

The two-shear theory of martensite formation was first confirmed by crystallographic measurements on the two phases, but the existence of the inhomogeneous lattice deformation could only be directly established by microscopic examination. Examination of a number of non-ferrous martensite transformations in the optical microscope revealed that the martensite lamellae contained numerous very fine twins in uniform arrays. For example, the martensite reaction in the indium-thallium system has some similar characteristics to ferrous martensites in so far as the transformation is from face-centred cubic to a tetragonal lattice (face-centred). The martensitic lamellae are very uniform, and contain fine twins on a single variant (101) $[10\bar{1}]$ in one lamella, followed by twinning in the opposite direction (101) $[\bar{1}01]$ in the next lamella and so on.

Martensitic plates in steel are frequently not parallel-sided, instead they are often lenticular as a result of constraints in the matrix which oppose the shape change resulting from the transformation. This is one of the reasons why it is difficult to identify precisely habit planes in ferrous martensite. However, it is not responsible for the irrational planes, but rather the scatter obtained in experiments. Another feature of higher carbon martensites is the burst phenomenon, in which one martensite plate nucleates a sequence of plates presumably as a result of stress concentrations set up when the first plate reaches an obstruction such as a grain boundary or another martensite plate (Fig. 5.7).

Perhaps the most striking advances in the structure of ferrous martensites occurred when thin-foil electron microscopy was first used on this problem. The two modes of plastic deformation needed for the inhomogeneous deformation part of the transformation, i.e. slip and twinning, were both observed by Kelly and Nutting. All ferrous martensites show very high dislocation densities of the order of 10^{11} to 10^{12} cm^{-2} (Fig. 5.10b), which are similar to those of very heavily cold-worked alloys. Thus it is usually impossible to analyse systematically the planes on which the dislocations occur or determine their Burgers vectors.

The lower carbon (<0.5%C) martensites on the whole exhibit only dislocations. At higher carbon levels very fine twins (5–10 nm wide) commonly occur (Fig. 5.11b). The twinning plane is $\{112\}_{\alpha'}$ derived from $\{110\}_{\gamma}$, and the twinning direction is $\langle 111 \rangle_{\alpha'}$ corresponding to the $\langle 110 \rangle_{\gamma}$ direction. In favourable circumstances the twins can be observed in the optical microscope, but the electron microscope allows the precise identification of twins by the use of the selected area electron diffraction technique. Thus the twin shears can be analysed precisely and have provided good evidence for the correctness of the crystallographic theories discussed above. However, twinning is not always fully developed and even within one plate some areas are often untwinned. The phenomenon is sensitive to composition.

The evidence suggests that deformation by dislocations and by twinning are alternative methods by which the lattice invariant deformation occurs. From general knowledge of the two deformation processes, the critical resolved shear stress for twinning is always much higher than that for slip on the usual slip plane. This applies to numerous alloys of different crystal structure. Thus it might be expected that those factors which raise the yield stress of the austenite, and martensite, will increase the likelihood of twinning. The important variables are: carbon concentration; alloying element concentration; temperature of transformation; strain rate.

The yield stress of both austenite and martensite increases with carbon content, so it would normally be expected that twinning would, therefore, be encouraged. Likewise, an increase in the substitutional solute concentration raises the strength and should also increase the incidence of twinning, even in the absence of carbon, which would account for the twins observed in martensite in high concentration binary alloys such as Fe-32%Ni. A decrease in transformation temperature, i.e. reduction in M_s, should also help the formation of twins, and one would particularly expect this in alloys transformed, for example, well below room temperature. It should also be noted that carbon concentration and alloying element concentration should assist by lowering M_s. As martensite forms over a range of temperatures, it might be expected in some steels that the first formed plates would be free of twins whereas the plates formed nearer to M_s would more likely be twinned. The observed inhomogeneities within plates could arise if growth of the plate near M_s was continued at lower temperatures. However, often plates have a mid-rib along which twinning occurs, the outer regions of the plate being twin-free. This could possibly take place when the M_s is below room temperature leading to twinned plates which might then grow further on resting at room temperature.

Returning to the three types of martensite referred to in 5.2, the morphological and crystallographic characteristics can now be summarized.

Low carbon martensite
    Habit plane $\{111\}_{\gamma}$
    Kurdjumov-Sachs relation $\{111\}_{\gamma} \{110\}_{\alpha'} \langle 011 \rangle_{\gamma} // \langle 111 \rangle_{\alpha'}$

Referred to as lath martensite

This type of martensite is found in plain carbon and low alloy steels up to about 0.5 wt% carbon. The morphology is lath- or plate-like (Fig. 5.10a), where the laths are very long and about 0.5 μm wide. These are grouped together in packets with low angle boundaries between each lath, although a minority of laths is separated by high angle boundaries (Fig. 5.10b). In plain carbon steels practically no twin-related laths have been detected.

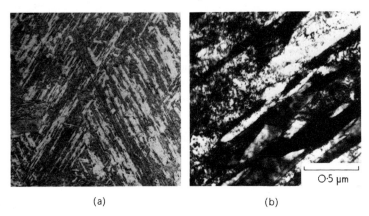

Fig. 5.10 Fe-0.16C alloy. Martensite formed by quenching from 1050°C: a, optical micrograph, ×95; b, thin-foil EM showing heavily dislocated laths (Ohmori)

However, in iron-nickel alloys adjacent laths are frequently twin-related. The boundaries between laths are not strictly planar, nor can they be described as lenticular, but dovetailing within the packets is frequent. Internally, the laths are highly dislocated and it is frequently difficult to resolve individual dislocations which form very tangled arrays. Twins are not observed to occur extensively in this type of martensite.

Medium carbon martensite

Habit plane $\{225\}_\gamma$

Kurdjumov-Sachs relation

Referred to as acicular

It is perhaps unfortunate that the term acicular is applied to this type of martensite because its characteristic morphology is that of lenticular plates (Fig. 5.11), a fact easily demonstrated by examination of plates intersecting two surfaces at right angles. These plates first start to form in steels with about 0.5% carbon (Fig. 5.12), and can be concurrent with lath martensite in the range 0.5%–1% carbon. Unlike the laths, the lenticular plates form in isolation rather than in packets, on planes approximating to $\{225\}_\gamma$ and on

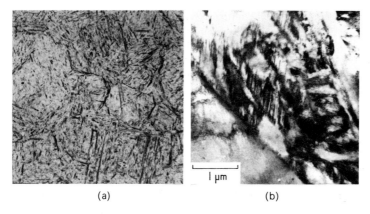

Fig. 5.11 Fe-0.8C alloy quenched from 1100°C: a, optical micrograph ×200; b, thin-foil EM showing twinning in martensite laths (Ohmori)

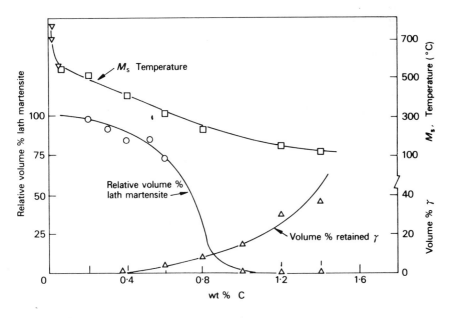

Fig. 5.12 Effect of carbon content on the type of martensite and the amount of retained austenite in Fe-C alloys (Speich, *Met. Trans.*, 1972, **3**, 1045)

several variants within one small region of a grain, with the result that the structure is very complex (Fig. 5.11). The burst phenomenon probably plays an important part in propagating the transformation, and the austenite is thus not as uniformly or as efficiently eliminated as with lath martensites. This physical difference cannot be unconnected with the fact that higher percentages of retained austenite occur as the carbon level is increased (Fig. 5.12), and the martensite is predominantly lenticular. The micro-twinning referred to earlier is found predominantly in this type of martensite (Fig. 5.11b), which forms at lower M_s temperatures, as the carbon content increases.

High carbon martensite
   Habit plane $\{259\}_\gamma$
   Nishiyama $\{111\}_\gamma // \{110\}_{\alpha'}$ $\langle 112 \rangle_\gamma // \langle 011 \rangle_{\alpha'}$

When the carbon content is >1.4 wt%, the orientation relationship changes from Kurdjumov-Sachs to Nishiyama, and the habit plane changes to around $\{259\}_\gamma$. The change is not detectable microscopically as the morphology is still lenticular plates which form individually and are heavily twinned. Detailed crystallographic analysis shows that this type of martensite obeys more closely the theoretical predictions than the $\{225\}$ martensite. The plates are formed by the burst mechanism and often an audible click is obtained (cf. mechanical twinning). The $\{259\}$ martensite only forms at very high carbon levels in plain carbon steels, although the addition of metallic alloying elements causes it to occur at much lower carbon contents, and in the extreme case in a carbon-free alloy such as Fe-Ni when the nickel content exceeds about 29 wt%.

5.6 Kinetics of transformation of martensite

Martensitic transformations are usually described as athermal, since transformation commences at a well-defined temperature M_s, but for transformation to continue the temperature must continue to fall until M_f is reached when the reaction is complete. However, there are martensitic reactions which can proceed at constant temperature.

5.6.1 Nucleation and growth of martensite

The driving force for the start of transformation can be expressed as $T_0 - M_s$ where T_0 is the temperature at which austenite and martensite possess the same free energy. Fig. 5.13 shows this schematically by plotting the two curves for the free energy of austenite and of martensite as a function of temperature. The M_s temperature is also shown as is the A_s temperature, the temperature at which martensite starts to revert to austenite on reheating. Both reactions require a degree of supercooling or superheating. Observations on numerous systems indicate that where the transformation results in a large shape change, the driving force $(T_0 - M_s)$ is large and the temperature range $M_s - M_f$ is also large, whereas with small shape changes the reverse is true. With ferrous martensites the shape

The formation of martensite

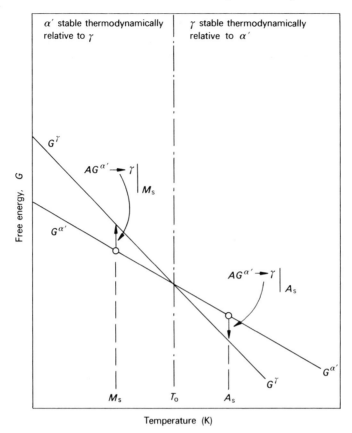

Fig. 5.13 Free-energy of austenite and martensite as a function of temperature (Kaufmann and Cohen, *Progress in Metal Physics*, 1958, **7**, 165)

change is large and the $M_s - M_f$ range is often several hundred degrees. It seems likely, therefore, that the strain energy arising when a small martensite plate is formed plays a significant role in nucleation.

The classical theory of homogeneous nucleation can be applied to an athermal reaction where either
(a) the nuclei form rapidly at M_s, or
(b) subcritical nuclei pre-exist which become supercritical at M_s.
The overall free energy change, ΔG, when nucleation takes place, is a result of three components:

the change in chemical free energy, Δg
the strain energy
the interfacial energy between matrix and martensite.

For a semicoherent nucleus of martensite with an oblate spheroid shape, radius r, semi-thickness c (Fig. 5.14):

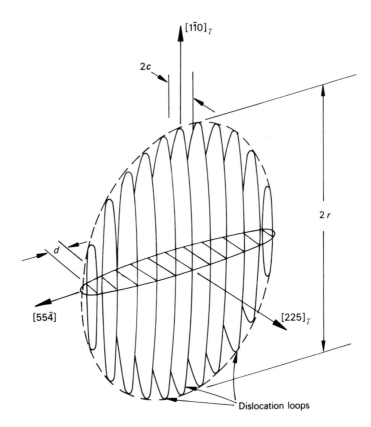

Fig. 5.14 Model of a martensitic nucleus (Knapp and Dehlinger, In: Kaufmann and Cohen, *Progress in Metal Physics*, 1958, **7**, 165)

$$\Delta G = \frac{4}{3}\pi r^2 c\Delta g + \frac{4}{3}\pi rc^2 A + 2\pi r^2 \alpha \tag{5.2}$$

where A = strain energy factor
α = free energy per unit area of γ/α' interface
Δg = chemical free energy change per unit volume.

The critical nucleus size (σ) is determined by c^* and r^*, at which the free energy change is ΔG^*, which defines a saddle on the free energy-c-r curve, thus:

$$c^* = -2\sigma/\Delta g \tag{5.3}$$
$$r^* = 4A\sigma/\Delta g^2 \tag{5.4}$$

and
$$\Delta G^* = 32\pi A^2 \sigma^3 / 3(\Delta g)^4 \tag{5.5}$$

However, if reasonable values of Δg, A and σ are used in equation (5.5), the value of ΔG^* is so high ($\simeq 3 \times 10^5$ kJ) that the barrier to nucleation is orders of magnitude too large. It would, therefore, be quite impossible for martensite nuclei to occur as a result of random fluctuations.

The results of these calculations suggest that nucleation of martensite must take place heterogeneously on pre-existing embryos, which it is assumed are already beyond the saddle point in the free energy curve. However, the search for such nuclei has not been very successful and they still remain a deduction from formal nucleation theory. In some special cases nuclei can be obtained. For example, in high manganese steels stacking faults readily occur as the austenite has a low stacking fault energy. On transformation to martensite, an ε-martensite of hexagonal structure is obtained which has been shown to nucleate at stacking faults.

The embryos are postulated to have a semicoherent dislocation interface with the austenite, envisaged as arrays of parallel dislocation loops which join the embryo to its matrix (Fig. 5.14). Growth then takes place by nucleation of new dislocation loops which join the interface and extend it. Recently, Olson and Cohen have developed a new theory of nucleation in which the first step is faulting on the closest packed planes derived from existing groups of dislocations. The most likely sites for such nuclei are grain boundaries, incoherent twin boundaries and inclusion particle interfaces.

Normally individual martensite plates grow at extremely rapid rates, forming in times of the order of 10^{-7} seconds. It has been found that the growth velocity is constant over a wide temperature range which indicates that the growth process is not thermally activated. This is consistent with the crystallographic evidence that the atomic movements are small and orderly, and that atoms do not change places by diffusion. The growth is envisaged as the movement of an array of parallel dislocations lying in the interface, all having the same Burgers vector. As the interface moves forward into the austenitic matrix the dislocations keep up with the interface by gliding on the appropriate slip planes. This type of movement involves motion of the habit plane in a direction normal to itself.

Isothermal growth of martensite plates has often been observed at rates permitting direct observation in the optical microscope, e.g. in iron-nickel-manganese alloys. Other alloys, e.g. iron-nickel and iron-nickel-carbon, exhibit the burst phenomenon, although there is substantial evidence that isothermal transformation often takes place in alloys with low M_s which exhibited this phenomenon. In these cases it seems that the main factor is slow isothermal nucleation rather than slow growth.

Looking at the kinetics of martensite formation in broad terms, there are three different types of behaviour which can take place (Fig. 5.15). The first

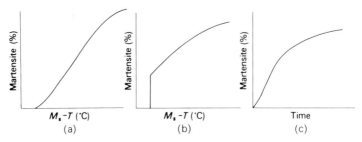

Fig. 5.15 Transformation curves for martensite: a, athermal transformation; b, athermal with bursts; c, isothermal transformation (Christian, In: *Martensite: Fundamentals and Technology*, ed. E. R. Petty, Longmans, 1970)

type involves normal athermal transformation with a sigmoidal type of curve where the fraction of austenite transformed is a function solely of the temperature (Fig. 5.15a). The second type also involves athermal transformation, but the reaction commences suddenly with a burst phenomenon which effectively causes a proportion of the austenite to transform isothermally (Fig. 5.15b). Further transformation is again athermal in character. Finally, with true isothermal transformation (Fig. 5.15c) the proportion of austenite transformed is proportional to time at a given temperature. This last type of behaviour has only been found in carbon-free iron base alloys.

5.6.2 Effect of alloying elements

Most alloying elements which enter into solid solution in austenite lower the M_s temperature, with the exception of cobalt and aluminium. However, the interstitial solutes carbon and nitrogen have a much larger effect than the metallic solutes. The effect of carbon on both M_s and M_f is shown in Fig. 5.16, from which it can be seen that 1 wt % of carbon lowers the M_f by over 300°C. Note that above 0.7%C the M_f temperature is below room temperature and consequently higher carbon steels quenched into water will normally contain substantial amounts of retained austenite.

The relative effect of other alloying elements is indicated in the following empirical relationship due to Andrews:

$$M_s(°C) = 539 - 423(\%C) - 30.4(\%Mn) - 17.7(\%Ni) - 12.1(\%Cr) - 7.5(\%Mo) \tag{5.6}$$

The effect of alloying elements on the austenite/martensite transformation was originally explained by a thermodynamic analysis due to Zener. Using a binary Fe-X system equations can be written for the chemical free energy of the austenite G^γ and martensite $G^{\alpha'}$ phases. In austenite

$$G^\gamma = (1-x)G^\gamma_{Fe} + xG^\gamma_X + G^\gamma_M \tag{5.7}$$

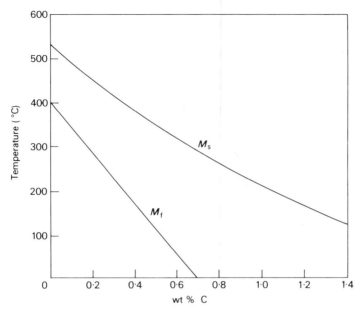

Fig. 5.16 The effect of carbon on M_s and M_f (Petty (ed.), *Martensite: Fundamentals and Technology*, Longmans, 1970)

where x is the atomic fraction of alloying element; G^γ_{Fe} is the free energy of iron in the γ form; G^γ_X is the free energy of element X in the γ form, which must be deduced for elements that do not exist in fcc form; and G^γ_M is the free energy of mixing of austenite.

A similar equation can be written for $G^{\alpha'}$ and subtracting from Equation (5.7) gives

$$\Delta G^{\alpha' \to \gamma} = (1-x)\Delta G^{\alpha \to \gamma}_{Fe} + x\Delta G^{\alpha \to \gamma}_X + \Delta G^{\alpha' \to \gamma}_M \tag{5.8}$$

Zener approached the alloying problem by assuming that the solid solutions were sufficiently dilute to be ideal, so that the mixing term $\Delta G^{\alpha' \to \gamma}_M$ is zero. Now,

$$\Delta G^{\alpha \to \gamma}_X = \Delta H^{\alpha \to \gamma}_X - T\Delta S^{\alpha \to \gamma}_X \tag{5.9}$$

where ΔS is the entropy change between α and γ, ΔH is the enthalpy change and T is the temperature. Also

$$\Delta G^{\alpha \to \gamma}_X = RT \ln \frac{x_\alpha}{x_\gamma} \tag{5.10}$$

where x_α and x_γ are the compositions of α and γ in equilibrium with γ and α at any temperature.

Zener simplified the argument by assuming that $RT \ln \frac{x_\alpha}{x_\gamma}$ is constant, and that $\Delta S_X^{\alpha \to \gamma}$ is zero, so that with ideal solutions

$$\Delta H_X^{\alpha \to \gamma} = RT\ln\frac{x_\alpha}{x_\gamma} = \Delta H_X^{\alpha' \to \gamma} \tag{5.11}$$

which is defined as the difference in enthalpies of alloying element X in the austenitic and martensitic phases. Therefore, Equation (5.8) can now be rewritten, expressing the driving force of the reaction $\Delta G^{\alpha' \to \gamma}$ as

$$\Delta G^{\alpha' \to \gamma} = (1-x)\Delta G_{Fe}^{\alpha \to \gamma} + x\Delta H_X^{\alpha' \to \gamma} \tag{5.12}$$

After determining $\Delta H_X^{\alpha' \to \gamma}$, the free energy change for the martensite reaction $\Delta G^{\alpha' \to \gamma}$ can be calculated. Values for T_o, the temperature at which γ and α' have the same free energies, can be calculated by putting $\Delta G^{\alpha' \to \gamma}$ equal to zero.

Alloying elements either expand the γ-loop, i.e. stabilize γ, or contract the loop and encourage α-formation, and this will have different effects on $\Delta H_X^{\alpha' \to \gamma}$ (Chapter 4, Section 1). Elements which expand the γ-loop will make this term negative and lower T_o, while elements which favour α-formation will make the term positive and raise T_o.

It is interesting to look at the values of $\Delta H_X^{\alpha' \to \gamma}$ which are available in the literature for a number of common alloying elements (Table 5.1). There are some anomalies, for example chromium, which contracts the γ-loop, has a negative ΔH value, suggesting that ΔH has been computed from data at too low a temperature.

Table 5.1 Values of difference in enthalpy of element X in austenite (γ) and in martensite (α')

X	C	N	Mn	Ni	Cu	Cr	W	Mo	V	Ti
$\Delta H_X^{\alpha' \to \gamma}$ (kJ mol^{-1})	−33.9	−22.4	−10.7	−8.4	−5.4	−5.0	+5.7	+5.7	+11.8	+37.7

Cohen and coworkers have provided detailed data for iron–carbon alloys between 0 and 1.1 wt% carbon and in Fig. 5.17 the temperature dependence of $\Delta G^{\alpha' \to \gamma}$ is plotted for several carbon levels. The intersection of the curves with the $\Delta G^{\alpha' \to \gamma} = 0$ axis provides values of T_o for the various compositions. It was found that the driving force $\Delta G^{\alpha' \to \gamma}$ at the M_s temperatures of the alloys was practically constant, approximately 1250 J mol^{-1}, independent of carbon content. However, work on iron–nickel alloys has shown that the driving force increases with increasing nickel content, i.e. as the M_s is depressed.

5.6.3 The effect of deformation

The effect of stress on the martensitic transformation is normally to raise the M_s temperature. The superimposed stress field from the plastic, or

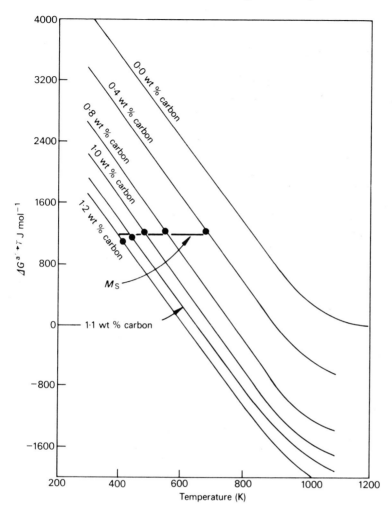

Fig. 5.17 Free energy change for the austenite-martensite reaction as a function of temperature and carbon content (Kaufmann and Cohen, *Progress in Metal Physics*, 1958, **7**, 165)

elastic, deformation reinforces that caused by the nucleation of a martensite plate, and in one sense the subsequent shape change is a further plastic deformation process. We can define a temperature M_d, greater than M_s, above which deformation of the parent phase does not form any martensite. However, it is likely that deformation of the austenite above M_d will alter the M_s on subsequent cooling through the martensitic range. Usually in these circumstances, the M_s is lowered, and the resultant increased stability of the austenite is referred to as *mechanical stabilization*.

5.6.4 Stabilization

Stabilization means a reduction in the amount of transformation of austenite to martensite, as a result of processes which interfere with the nucleation and growth of the plates. Plastic deformation above the M_d temperature can achieve this. However, the term stabilization is normally applied when the cooling of a steel is arrested in the $M_s - M_f$ range. The transformation, when resumed by lowering the temperature, does not result in as complete a transformation to austenite as would have been the case if no isothermal pause had occurred. At the chosen delay temperature, the degree of stabilization increases to a maximum with time, and as the temperature approaches M_f, the extent of stabilization increases. It appears that stabilization is at a minimum when only a small amount of martensite is present in the matrix.

The explanation of these complex effects lies in the fact that the formation of martensite plates leads to accommodating plastic deformation in the surrounding matrix, which can result in high concentrations of dislocations in the austenite. Interaction of some of these dislocations with the glissile dislocations in the martensite plate boundary will then cause it to be no longer mobile, so that the plate cannot grow further. Any phenomena which help to encourage this process will achieve stabilization. Resting at an intermediate temperature gives time for plastic relaxation, i.e. movement of dislocations, as well as the locking of interfacial dislocations by carbon atoms.

5.7 The strength of martensite

The high hardness and brittleness of rapidly quenched steels is the result of the formation of martensite, yet many shear transformations in non-ferrous alloy systems do not produce this dramatic hardening. Indeed, if carbon is eliminated from the steel the resulting hardness is very much lower. Fig. 5.18 shows the large effect of carbon content on the hardness of martensite compared with the relatively small effect of carbon on the strength of austenite, retained to room temperature by the addition of nickel.

The strength levels reached depend also on the detailed structure of the martensite, e.g. whether it has remained stable during quenching and testing at room temperature. By addition of nickel to iron carbon alloys, Winchell and Cohen depressed the M_s temperature to $-35°C$, so that martensite formed only at low temperatures and auto-tempering was eliminated (Chapter 8). In addition, the samples were deformed at $0°C$, with the results shown in Fig. 5.19, indicating that the flow stress of martensite increases with carbon content up to about 0.5 wt %C. Allowing the martensite to rest for 3 hours at $0°C$, resulted in the upper curve (Fig. 5.19), demonstrating that martensite can age harden at ambient temperature or below.

The question of the origin of the high strength of martensite is a difficult

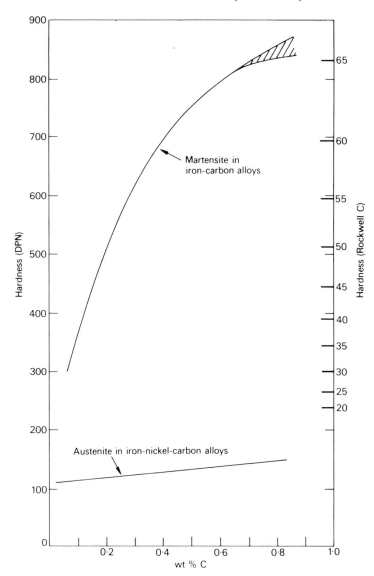

Fig. 5.18 The effect of carbon on the hardness of martensite and austenite (Winchell and Cohen, *Trans. Met. Soc. AIME*, 1962, **224**, 638)

one, compounded by the complexity of the structure, a tetragonal lattice with interstitial carbon in solid solution, formed by shear which leads to high densities of dislocations and fine twins. There are, as a result, several possible strengthening mechanisms:
(1) substitutional and interstitial solid solution

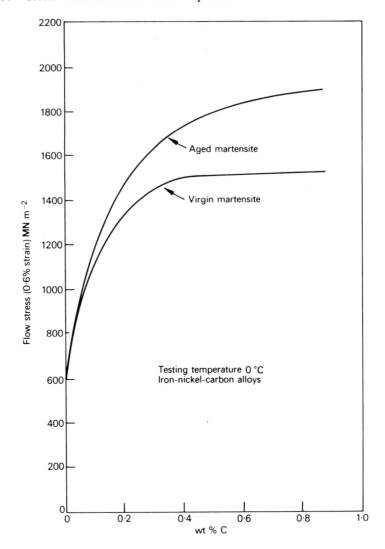

Fig. 5.19 Ageing of martensite at 0°C in Fe-Ni-C alloys (Winchell and Cohen, *Trans. Met. Soc. AIME*, 1962, **224**, 638)

(2) dislocation strengthening, i.e. work hardening
(3) fine twins
(4) grain size
(5) segregation of carbon atoms
(6) precipitation of iron carbides

The interstitial solid solution of carbon which results in the tetragonality of martensite is a prime candidate for the role of major strengthening factor.

The work of Winchell and Cohen enabled the determination of the yield stress as a function of carbon content under conditions when the carbon atoms were unable to diffuse to form atmospheres and precipitates. The flow stress was shown to vary as $c^{\frac{1}{3}}$, where c = carbon concentration, but later it was found that the strengh could be shown equally well to vary as $c^{\frac{1}{2}}$.

Fleischer examined the situation theoretically with a model of a dislocation bending away from interstitial solute atoms with short range interactions, and using a parameter $\Delta\varepsilon$, the difference in longitudinal and transverse lattice strain caused by an interstitial carbon atom in martensite ($\Delta\varepsilon \simeq 0.38$). He found the following expression for the flow stress τ:

$$\tau = \tau_0 + \frac{2G\Delta\varepsilon c^{\frac{1}{2}}}{3} \tag{5.13}$$

predicting that the flow stress is proportional to $c^{\frac{1}{2}}$. The curve has a slope of $\frac{G}{15}$ to $\frac{G}{20}$. Other experiments on martensites with low M_s temperatures support the $c^{\frac{1}{2}}$ relationship, with slight differences in slope depending on whether the martensite is of lath type or twinned (Fig. 5.20).

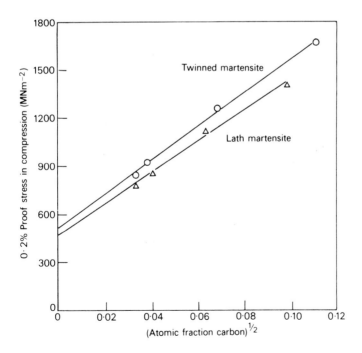

Fig. 5.20 Effect of carbon on the strength of martensite (Chilton and Kelly, *Acta Met.*, 1968, **16**, 637)

The proposal that the fine twins characteristic of higher carbon martensites makes a major contribution to strength has not received wide acceptance. Certainly, a large increase in strength is not found when the transition from dislocated martensite to twinned martensite takes place. However, the high dislocation densities of twin-free martensite must make some contribution to strength, estimated to be not greater than 300 MNm^{-2}, and there is reason to believe that the fine-twinning makes a similar, but not additive, contribution.

The austenitic grain size determines the maximum size of a martensitic plate, so some dependence of strength on grain size might be expected. In fact, a Petch-type plot has been found for several alloy steels of different austenitic grain sizes tested in the martensitic condition (Fig. 5.21). However, when the fine structure of martensite is examined other possible grain sizes much finer than the austenitic grain size can be considered as

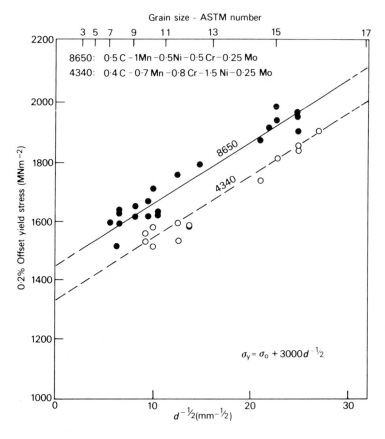

Fig. 5.21 The effect of prior austenite grain size on the strength of martensite (Grange, *Trans. ASM*, 1966, **59**, 26)

contributors to strength. Firstly, there is the packet size in lath martensite, or the individual plate in lenticular martensite, and beyond these there is the lath substructure which is usually well below 1 μm in thickness. While many of these boundaries are really low angle subboundaries, they do present obstacles to dislocation movement and must, therefore, be considered to make some contribution to the overall strength.

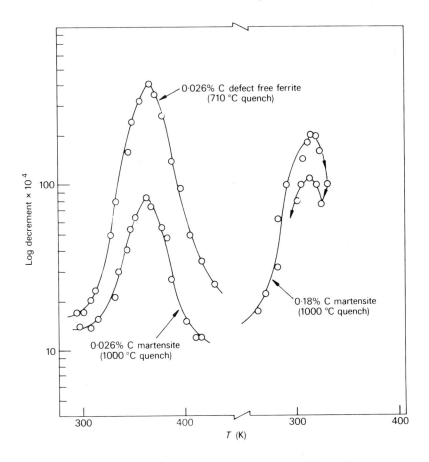

Fig. 5.22 Comparison of the internal friction behaviour of low carbon martensites with that of ferrite (Speich, *Trans. Met. Soc. AIME*, 1969, **245**, 2553)

It is also to be expected that carbon atoms are segregated to the high dislocation population in martensite, bearing in mind the strong interactions found in ferrite. Kurdjumov and coworkers have used the internal friction technique to study this problem by examining the amplitude dependence of the internal friction after allowing martensite to rest at room temperature. Dislocation damping is the principal contribution to internal friction observed at low temperatures, which is characterized by the presence of a well-defined peak, the Snoek peak. Kurdjumov showed that this peak was much lower in 0.026%C martensite, than in ferrite of the same carbon level quenched from 710°C (Fig. 5.22). The ferrite has a very low dislocation population relative to that of the martensite, yet it exhibits the highest peak. This indicates that the dislocations in the martensite are effectively pinned by carbon atoms, consequently these dislocations do not contribute effectively to the damping process.

Work on the temperature dependence of the flow stress of martensite in Fe-Ni-C alloys has shown a strong temperature dependence, together with a peak in the curve associated with serrated flow in the stress-strain curve (Fig. 5.23). Like the development of the yield point in α-iron, this has been attributed to the Cottrell-Bilby interaction of carbon with dislocations.

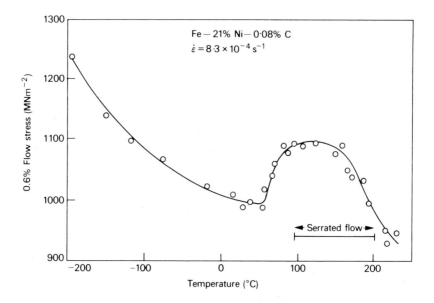

Fig. 5.23 Temperature dependence of the flow stress of Fe-Ni-C martensite (Owen and Roberts, *International Conference on Strength of Metals and Alloys*, Tokyo, 1968)

However, this phenomenon leads to precipitation of iron carbide on the dislocations which is responsible for the increase in strength shown by martensite aged at room temperature or just above. Also martensites with relatively high M_s temperatures will form cementite dispersions during the quench (auto-tempering) which will also make some contribution to the observed strength.

The yield strength of martensite, like that of ferrite, is markedly temperature dependent, but this dependence is little affected by the presence or absence of precipitate or by the amount of carbon in solution. It is, therefore, likely that the temperature dependence arises from the basic resistance of the lattice to dislocation movement, i.e. it is a result of the temperature dependence of the Peierls-Nabarro force.

Further reading

Kaufman, L. and Cohen, M., Thermodynamics and kinetics of martensite transformations. *Prog. in Metal Physics*, **7**, 165, 1958

Kurdjumov, G. V., Phenomena occurring in the quenching and tempering of steel, 12th Hatfield Memorial Lecture. *J.I.S.I.*, **195**, 26, 1960

Cohen, M., The strengthening of steel, 1962 Howe Memorial Lecture. *Trans. AIME*, **224**, 638, 1962.

Iron and Steel Institute, Physical Properties of Martensite and Bainite, Special Report No. 93, 1965.

Wayman, C. M., The crystallography of martensitic transformations, in alloys of iron. *Advances in Materials Research*, **3**, 147, 1968.

American Society for Metals, *Phase Transformations*, 1970

Petty, E. R. (ed), *Martensite, Fundamentals and Technology*, Longmans, 1970

Christian, J. W., In: *The Strength of Martensite in Strengthening Methods in Crystals*, ed. A. Kelly and R. B. Nicholson, Elsevier, 1971

Thomas, G., Electron microscopy investigations of ferrous martensites. *Met. Trans.*, **2**, 2373, 1971

Metallurgical Forum: Martensite, *J. Aust. Inst. Metals*, **19**, 3–60, 1974

Olsen, G. B. and Cohen, M., A general mechanism of martensitic nucleation, Parts I–III. *Met. Trans.*, **7A**, 1897–1923, 1976

AIME, Symposium on the formation of martensite in iron alloys, Las Vegas 1970. *Met. Trans.*, **2**, 2327–2462, 1971.

Nishiyama, Zenji, *Martensite Transformation*, English edition, Academic Press, New York, 1978

6
The bainite reaction

6.1 Introduction

Examination of the *TTT* diagram for a eutectoid carbon steel (Fig. 3.5), bearing in mind the fact that the pearlite reaction is essentially a high temperature one occurring between 550°C and 720°C and that the formation of martensite is a low temperature reaction, reveals that there is a wide range of temperature, usually 250–550°C, when neither of these phases forms. This is the region in which lath-shaped fine aggregates of ferrite and cementite are formed which possess some of the properties of the high temperature reactions involving ferrite and pearlite as well as some of the characteristics of the martensite reaction. The generic term for these intermediate structures is *bainite*, after Edgar Bain who with Davenport first found them during their pioneer systematic studies of the isothermal decomposition of austenite. Bainite also occurs during athermal treatments at cooling rates too fast for pearlite to form, yet not rapid enough to produce martensite.

The nature of bainite changes as the transformation temperature is lowered. Two main forms can be identified, *upper* and *lower bainite*.

6.2 Morphology and crystallography of upper bainite (temperature range 550–400°C)

The morphology of upper bainite bears a close resemblance to Widmanstätten ferrite, as it is composed of long ferrite laths free from internal precipitation. Fig. 6.1a shows upper bainite in a 0.8C steel at low magnification. Two-surface optical micrography decisively reveals that the ferrite component of upper bainite is composed of groups of thin parallel laths with a well-defined crystallographic habit (Fig. 6.1b). Like Widmanstätten ferrite, the bainitic ferrite laths exhibit the Kurdjumov-Sachs relationship with the parent austenite, but the relationship is less precise as the transformation temperature is lowered. A widely-accepted view is that the crystallography of upper bainite is very similar to that of low-carbon lath martensite. However, a detailed examination of the crystallography reveals that there are significant differences, and that upper bainite ferrite formation cannot be understood in terms of the crystallographic theory of martensite formation.

The bainite reaction 107

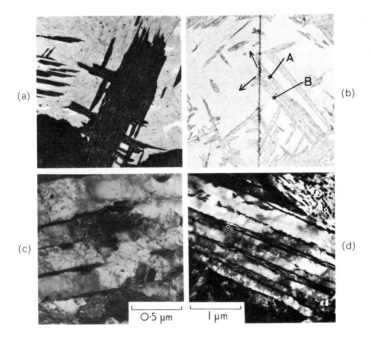

Fig. 6.1 Microstructure of upper bainite: a, 0.8C steel transformed 20s at 400°C, ×670; b, 0.34C steel transformed at 400°C. Two-surface composite micrograph, ×330; c, 0.8C steel transformed at 450°C. Dislocations visible. Thin-foil EM; d, 0.8C steel transformed 5s at 450°C. Thin-foil EM, (Ohmori)

Electron microscopy shows that upper bainite laths have a fine structure comprising smaller laths about 0.5 μm wide (Fig. 6.1c). These laths all possess the same variant of the Kurdjumov-Sachs relationship, so they are only slightly disoriented from each other. The longitudinal boundaries are, therefore, low angle boundaries. A typical austenite grain will have numerous sheaves of bainitic ferrite exhibiting the several variants of the Kurdjumov-Sachs orientation relationship, so large angle boundaries will occur between sheaves. The dislocation density of the laths increases with decreasing transformation temperature, but even at the highest transformation temperatures the density is greater than that in Widmanstätten ferrite.

The upper bainitic ferrite has a much lower carbon concentration ($<0.03\%$C) than the austenite from which it forms, consequently as the bainitic laths grow, the remaining austenite is enriched in carbon. This is an essential feature of upper bainite which forms in the range 550–400°C when the diffusivity of carbon is still high enough to allow partition between ferrite and austenite. Consequently, carbide precipitation does not occur within the laths, but in the austenite at the lath boundaries when a critical carbon concentration is reached.

The morphology of the cementite formed at the lath boundaries is dependent on the carbon content of the steel. In low carbon steels, the carbide will be present as discontinuous stringers and isolated particles along the lath boundaries, while at higher carbon levels the stringers may become continuous (Fig. 6.1d). With some steels, the enriched austenite does not precipitate carbide, but remains as a film of *retained austenite*. Alternatively, on cooling it may transform to high carbon martensite with an adverse effect on the ductility. This type of bainite is often referred to as *granular bainite*.

While there is no unique relationship between cementite and ferrite in upper bainite, the various observed relationships are compatible with the ferrite having a Kurdjumov-Sachs relation with the austenite and with the cementite being related to the austenite by the Pitsch relationship:

$$(001)_{Fe_3C} // (\bar{2}25)_\gamma$$

$$[100]_{Fe_3C} // [\bar{5}54]_\gamma$$

$$[010]_{Fe_3C} // [1\bar{1}0]_\gamma$$

The mode of growth of upper bainite laths has been the subject of considerable debate and, although there is some evidence of surface relief effects, it cannot be said that each lath forms as a result of a single tilt, as in the case of a martensite plate. Indeed, there have been studies of growth of upper bainite plates using thermionic emission microscopy, in which it has been demonstrated that movement of small ledges account for the thickening of plates, in a similar manner to that observed in Widmanstätten ferrite.

6.3 Morphology and crystallography of lower bainite (temperature range 400–250° C)

Lower bainite appears more acicular than upper bainite, with more clearly defined individual plates adopting a lenticular habit (Fig. 6.2a). Viewed on a single surface they misleadingly suggest an acicular morphology. However, two-surface optical microscopy of lower bainite indicates that the ferrite plates are much broader than in upper bainite, and closer in morphology to martensite plates (Fig. 6.2b). While these plates nucleate at austenitic grain boundaries, there is also much nucleation within the grains, i.e. *intragranular nucleation*, and secondary plates form from primary plates away from the grain boundaries.

Electron microscopy shows that the plates have a similar lath substructure to upper bainite, with the ferrite subunits about 0.5 μm wide and slightly disoriented from each other. The plates possess a higher dislocation density than upper bainite, but not as dense as in martensites of similar

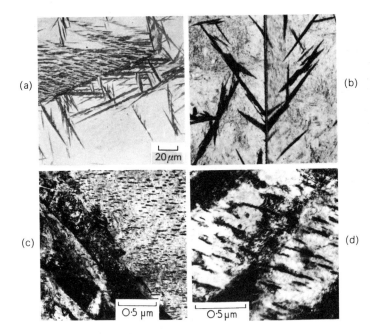

Fig. 6.2 Microstructure of lower bainite a, 0.8C steel transformed 30s at 300°C, ×260; b, 0.8C steel transformed at 300°C. Two surface composite micrograph, ×260; c, 0.8C steel transformed 6 min at 250°C. Thin-foil EM; d, 0.6C steel transformed at 300°C. Thin foil EM, (Ohmori)

composition. The crystallography of the plates seems to depend both on the temperature of transformation, and on the carbon content of the steel. Bowles and Kennon showed that, while at 400°C the habit plane was near $\{111\}_\alpha$, after transformation as low as 100°C the habit plane was much closer to $\{110\}_\alpha$, (Fig. 6.3). Moreover they showed that the phenomenological theory of martensite could be used for lower bainite to give satisfactory agreement between theory and experiment. Ohmori and coworkers have found that, in a 0.1 %C steel, bainite formed near the M_s has a $\{011\}_\alpha$ habit plane and a $\langle 111 \rangle_\alpha$ growth direction similar in behaviour to low carbon lath martensite. However, on increasing the carbon to 0.6–0.8 %C, the habit of the bainitic ferrite plates changes to $\{122\}_\alpha // \{496\}_\gamma$, which is not the same as for martensite of the same composition which has a $\{225\}_\gamma$ habit plane. Because of such variations, it has been suggested that lower bainite is not a true martensitic reaction. However, there is no reason to expect the transformations to be identical, and anyway the inhomogeneous shear would be expected to occur by slip in lower bainite, whereas twinning is the mode adopted in higher carbon martensites.

A striking microscopic characteristic of lower bainite is the growth of

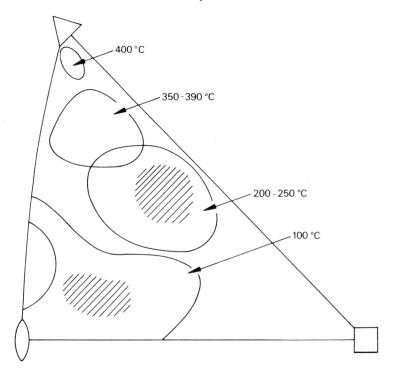

Fig. 6.3 Change of bainite habit plane with temperature of formation (Bowles and Kennon, J. Aust. Inst. Metals, 1960, **5**, 106)

carbide rods within the ferrite plates. The carbide can be either cementite or ε-iron carbide depending both on the transformation temperature, and on the composition of the steel, e.g. silicon encourages the formation of ε-iron carbide by retarding the nucleation of cementite. The cementite rods or laths have a $\langle 111 \rangle_\alpha$ growth direction, and an orientation relationship with the ferrite identical to that found in tempered martensite (Bagaryatski):

$[100]_C // [\bar{1}01]_\alpha$
$[010]_C // [111]_\alpha$
$[001]_C // [\bar{1}2\bar{1}]_\alpha$

However, in contrast to tempered martensite, the cementite particles in lower bainite exhibit only one variant of the orientation relationship, such that they form parallel arrays at about 60° to the axis of the bainite plate (Fig. 6.2c, d). This feature of the precipitate suggests strongly that it has not precipitated within plates supersaturated with respect to carbon, but that it has nucleated at the γ/α interface and grown as the interface has moved forward. It thus appears that the lower bainite reaction is basically an interface-controlled process leading to cementite precipitation, which then decreases the carbon content of the austenite and enhances the driving force for further transformation.

6.4 Reaction kinetics of bainite formation

In plain carbon steels, it is often difficult to separate the bainite reaction from the ferrite and pearlite reactions, as these phases can form under similar conditions to bainite. For example, the *TTT* diagram for a 0.8%C steel is a continuous curve although there is both a pearlite and bainite reaction occurring, but it is difficult to disentangle the reactions sufficiently to study their kinetics. However, the addition of certain alloying elements separates the reactions to the extent that they can be represented as individual curves on the *TTT* diagram, which then takes on a more complex form than the familiar C-curve, Fig. 6.4 is a *TTT* diagram for a 3%Cr 0.5C steel in which the pearlite and bainite reactions are very clearly separated, and the relationship of the bainite transformation to the M_s temperature is revealed.

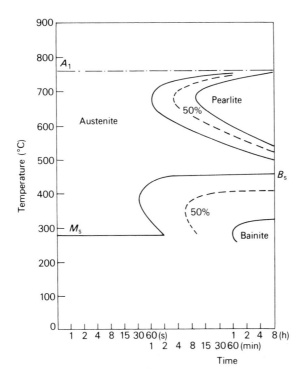

Fig. 6.4 *TTT* curves for 3%Cr 0.5C steel (Thelning, Steel and its Heat Treatment, *Bofors Handbook*, Butterworths, 1975)

There are two important features of bainite kinetics which can be shown by a variety of techniques, e.g. dilatometry, electrical resistivity, magnetic measurements and by metallography. Firstly, there is a well defined temperature B_s above which no bainite will form, which has been confirmed for a wide range of alloy steels and has been correlated with the structural transition from Widmanstätten ferrite to upper bainitic laths. Secondly, below B_s a temperature and time dependent process takes place over a wide temperature range (up to 150°C) which does not go to completion. Steels held in this range of temperature will not transform fully even after several months, but the residual austenite will in many cases transform during cooling to room temperature. A temperature B_f can also be defined, below which the bainite reaction will go to completion during isothermal transformation. However, this temperature appears not to have any fundamental significance.

The bainitic reaction has several of the basic features of a nucleation and growth process. It takes place isothermally, starting with an incubation period during which no transformation occurs, followed by an increasing rate of transformation to a maximum and then a gradual slowing down. These features are illustrated in the dilatometric results of Fig. 6.5 for three transformation temperatures in the bainitic range for a 1%Cr 0.4%C steel, the extent of transformation increasing with decreasing temperature. In this steel at 510°C the reaction stops after about one hour and the remaining austenite is stable at this temperature for a long time.

Using techniques such as thermionic emission microscopy it has been possible to study directly the progress of the bainite reaction. It has been found that upper bainitic plates lengthen and thicken during transformation by the movement along the plate boundaries of small steps which appear to be diffusion-controlled. The plates grow at a constant rate edgewise, which leads to a model for the reaction in which the driving force is provided by partition of the carbon from the ferrite to the austenite, the actual growth rate being determined by the rate of diffusion of carbon in austenite away from the γ/α interface. Some measurements have also been made on the thickening of upper bainite plates (Fig. 6.6) in which discontinuities of about 0.5 μm were observed resulting from the movement of steps along the interface. These step heights closely correlate with the size of subgrain units observed in upper bainite by thin foil electron microscopy (Fig. 6.1). This work seems to reduce the possibility that upper bainite is in any way formed by a martensitic reaction involving an invariant plane strain deformation. However the matter is still controversial.

In lower bainite there is more likelihood that a martensitic-type reaction occurs as the surface displacements observed appear to be uniform for individual plates, suggesting that they form from a shear-type reaction. Srinivasan and Wayman have made a detailed study of the lower bainite reaction in an Fe-8%Cr-1%C steel, in which lower bainite forms isothermally below 300°C, giving typical C-curve kinetics, but does not proceed

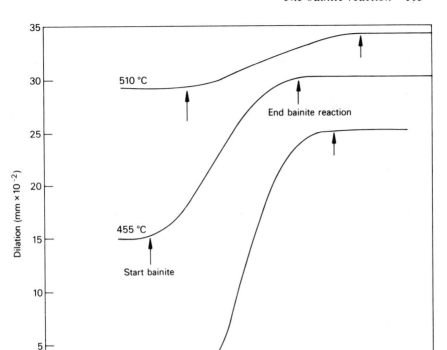

Fig. 6.5 Isothermal reaction curves for the formation of bainite in a steel with 1% Cr, 0.4%C (Hehemann, In: *Phase Transformations*, ASM, 1970)

to completion. An analysis of surface tilts from the initially small bainite plates led to the conclusion that the reaction was of the invariant plane strain-type characteristic of the martensite reaction, with an irrational bainitic habit plane near $\{254\}_\alpha$. However, the bainite plates do grow with time and new plates nucleate, hence the observed C-curve kinetics. As the thickening takes place the relief effects are more complex, and the simultaneous formation of iron carbide also complicates the reaction. It is, however, safe to say that the lower bainite reaction is partly martensitic in character, but possesses some of the characteristics of a diffusion-controlled process, as would be expected from the close association of iron carbide with the ferritic plates.

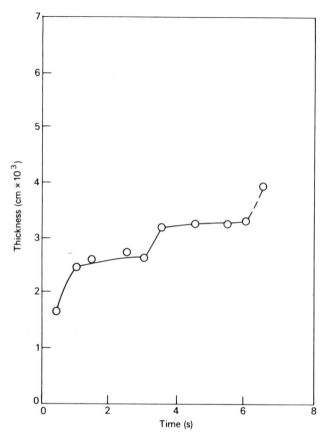

Fig. 6.6 Thickening kinetics of an upper bainite lath in an Fe-6Mn-1Mo-1C alloy transformed at 370°C (Aaronson, *Met. Trans.*, 1972, **3**, 1084)

6.5 Role of alloying elements

Carbon Carbon has a large effect on the range of temperature over which upper and lower bainite occur. The B_s temperature is depressed by many alloying elements but carbon has the greatest influence, as indicated by the following empirical equation:

$$B_S(°C) = 830 - 270(\%C) - 90(\%Mn) - 37(\%Ni) - 70(\%Cr) - 83(\%Mo) \quad (6.1)$$

The carbon concentration also influences the temperature of transition from upper to lower bainite in a rather complex way (Fig. 6.7). This transition temperature can be defined as the temperature at which diffusivity of carbon in austenite becomes too slow to allow diffusion away

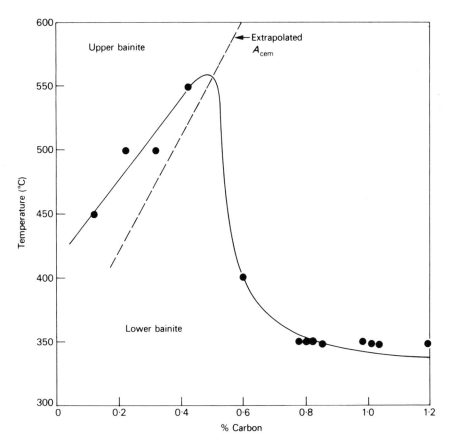

Fig. 6.7 Effect of carbon content on the temperature for transition from upper to lower bainite (Honeycombe and Pickering, *Met. Trans.*, 1972, **3**, 1099)

from the γ/α interface. Thus, in order that the growth of the bainitic ferrite can continue, precipitation of iron carbide must take place at the interface. In Fig. 6.8 the hypothetical carbon gradients near the γ/α interface are shown for a high and a low carbon steel. The higher carbon steel exhibits the shallower gradient which will be less efficient in moving carbon from the interface. Consequently, in higher carbon steels the transition from upper to lower bainite will occur at a higher temperature than in low carbon steels as shown in the left hand side of Fig. 6.7. However, above about 0.5%C, cementite would be expected to form directly from the supersaturated austenite, if the A_{cem} line on the Fe-C equilibrium diagram is extrapolated. This would allow the depleted austenite in the vicinity of the cementite to transform to upper bainite causing a sharp decrease in the temperature of transition from upper to lower bainite, as indicated in Fig. 6.7.

116 *Steels—Microstructure and Properties*

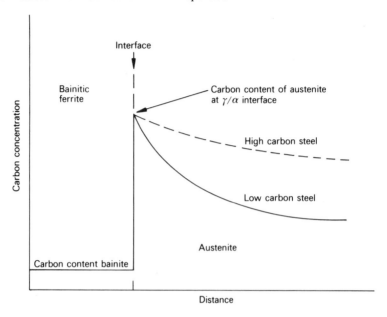

Fig. 6.8 Carbon concentration gradients during bainite formation (Pickering, In: *Transformation and Hardenability in Steels*, Climax Molybdenum Co., 1967)

Other alloying elements In plain carbon steels, the bainitic reaction is kinetically shielded by the ferrite and pearlite reactions which commence at higher temperatures and shorter times (Fig. 6.9a), so that in continuously-cooled samples bainitic structures are difficult to obtain. Even using isothermal transformation difficulties arise if, for example, the ferrite reaction is particularly rapid. As explained in Chapter 4, the addition of metallic alloying elements usually results in retardation of the ferrite and pearlite reactions. In addition, the bainite reaction is depressed to lower temperatures. This often leads to greater separation of the reactions, and the *TTT* curves for many alloy steels show much more clearly separate C-curves for the pearlite and bainitic reactions (Fig. 6.9b). However, it is still difficult to obtain a fully bainitic structure because of its proximity to the martensite reaction.

A very effective means of isolating the bainite reaction in low carbon steels has been found by adding about 0.002% soluble boron to a $\frac{1}{2}$% Mo steel. While the straight molybdenum steel encourages the bainite reaction (Fig. 6.9c), the boron markedly retards the ferrite reaction, probably by preferential segregation to the prior austenite boundaries. This permits the bainite reaction to occur at shorter times. Consequently, by use of a range of cooling rates, fully bainitic steels can be obtained.

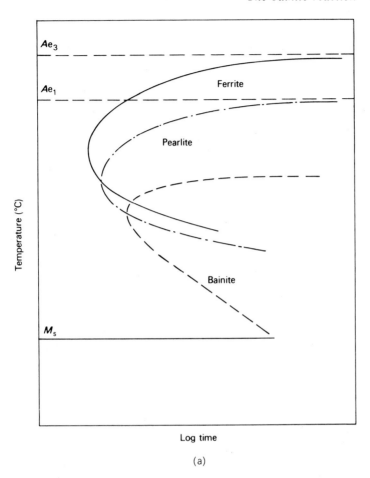

Fig. 6.9 Effect of alloying elements in the bainite reaction *TTT* curves: a, schematic diagram of a low carbon steel (Irvine and Pickering, *JISI*, 1957, **187**, 292)

6.6 Use of bainitic steels

Bainite frequently occurs in alloy steels during quenching ostensibly to form martensite. The cooling rate towards the centre of a steel bar is lower than the outside, so in large sections bainite can form in the inner regions with martensite predominating towards the surface. However, low carbon fully bainitic steels have been developed, as described, using $\frac{1}{2}$%Mo and very small concentrations of boron, which allow bainite to form over a wide range of cooling rates. Further control of the reaction is obtained by use of metallic alloying elements such as Ni, Cr, Mn which depress the tempera-

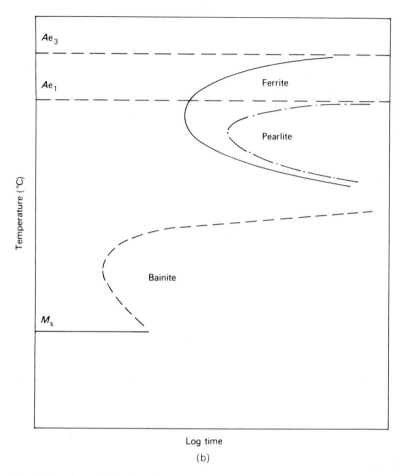

Fig. 6.9 Effect of alloying elements in the bainite reaction. *TTT* curves: b, schematic diagram of a low alloy steel (Irvine and Pickering, *JISI*, 1957, **187**, 292)

ture of maximum rate of formation of bainite. As the transformation temperature is lowered, for a constant cooling rate, the strength of the steel increases substantially. For a series of steels with 0.2% C the tensile strength can be varied between 600 and 1200 MNm^{-2}. However, this increase in strength is accompanied by a loss of ductility.

The practical advantage of the steels described above is that relatively high strength levels together with adequate ductility can be obtained without further heat treatment, after the bainite reaction has taken place. The steels are readily weldable, because bainite rather than martensite will form in the heat-affected zone (HAZ) adjacent to the weld metal, and so the

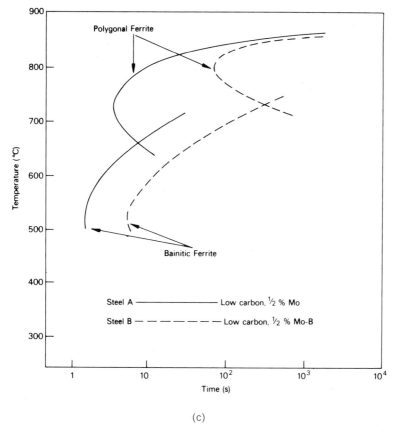

(c)

Fig. 6.9 Effect of alloying elements in the bainite reaction. TTT curves: c, low carbon 0.5Mo, with and without B (Irvine and Pickering, *J/S/*, 1957, **187**, 292)

incidence of weld cracking will be reduced. Furthermore, the steels have low carbon contents, which improves the weldability and reduces stresses arising from the transformation.

Further reading

Irvine, K. J. and Pickering, F. B., Low-carbon bainitic steels, *J.I.S.I.* **187**, 292, 1957

Iron and Steel Institute, *Physical Properties of Martensite and Bainite*, Special Report No. 93, 1965

Transformation and Hardenability in Steels, Symposium published by Climax Molybdenum, 1968.

American Society for Metals, *Phase Transformations*, 1970

Edgar Bain 80th Birthday Seminar, *Met. Trans.*, **3**, 1031–1121, 1972
Metallurgical Forum: Martensite, *J. Aust. Inst. Metals*, **19**, 3–60, 1974
Pickering, F. B., *Physical Metallurgy and the Design of Steel*, Applied Science Publishers, 1978

7
The heat treatment of steels–hardenability

7.1 Introduction

The traditional route to high strength in steels is by quenching to form martensite which is subsequently reheated or *tempered* at an intermediate temperature, increasing the toughness of the steel without too great a loss in strength. Therefore, for the optimum development of strength, a steel must first be fully converted to martensite. To achieve this, the steel must be quenched at a rate sufficiently rapid to avoid the decomposition of austenite during cooling to such products as ferrite, pearlite and bainite. The effectiveness of the quench will depend primarily on two factors: the geometry of the specimen, and the composition of the steel.

A large diameter rod quenched in a particular medium will obviously cool more slowly than a small diameter rod given a similar treatment. Therefore, the small rod is more likely to become fully martensitic. It has already been shown that the addition of alloying elements to a steel usually move the *TTT* curve to longer times, thus making it easier to pass the nose of the curve during a quenching operation, i.e. the presence of alloying elements reduces the critical rate of cooling needed to make a steel specimen fully martensitic. If this critical cooling rate is not achieved a steel rod will be martensitic in the outer regions which cool faster but, in the core, the slower cooling rate will give rise to bainite, ferrite and pearlite depending on the exact circumstances.

The ability of a steel to form martensite on quenching is referred to as the *hardenability*. This can be simply expressed for steel rods of standard size, as the distance below the surface at which there is 50% transformation to martensite after a standard quenching treatment, and is thus a measure of the depth of hardening.

7.2 Use of *TTT* and continuous cooling diagrams

TTT diagrams provide a good starting point for an examination of hardenability, but as they are statements of the kinetics of transformation of austenite carried out *isothermally*, they can only be a rough guide. To take one example, the effect of increasing molybdenum, Fig. 7.1 shows the *TTT* diagrams for a 0.4%C 0.2% Mo steel and a steel with 0.3%C 2% Mo. The 0.2% Mo steel begins to transform in about one second at 550°C, but on

122 Steels—Microstructure and Properties

increasing the molybdenum to 2% the whole C-curve is raised and the reaction substantially slowed so that the nose is above 700°C, the reaction starting after 4 minutes. The latter steel will clearly have a greatly enhanced hardenability over that of the 0.2 Mo steel.

The obvious limitations of using isothermal diagrams for situations involving a range of cooling rates through the transformation temperature range have led to efforts to develop more realistic diagrams, i.e. continuous cooling (CCT) diagrams. These diagrams record the progress of the transformation with falling temperature for a series of cooling rates. They are determined using cylindrical rods which are subjected to different rates of cooling, and the onset of transformation is detected by dilatometry, magnetic permeability or some other physical technique. The products of

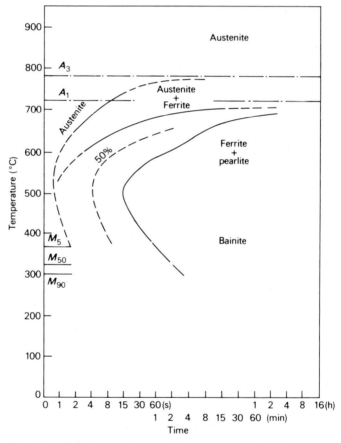

Fig. 7.1a TTT diagram of a molybdenum steel: 0.4C, 0.2Mo (Thelning, *Steel and its Heat Treatment*, Bofors Handbook, Butterworths, 1975)

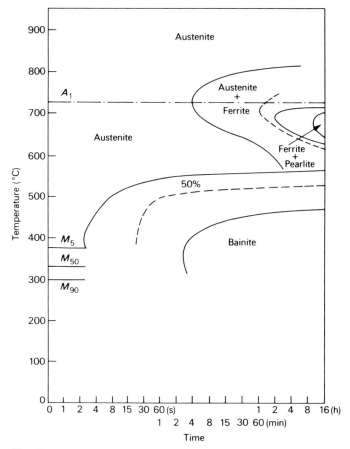

Fig. 7.1b *TTT* diagram of a molybdenum steel: 0.3C, 2.0Mo (Thelning, Steel and its Heat Treatment, *Bofors Handbook*, Butterworths, 1975)

the transformation, whether ferrite, pearlite or bainite, are partly determined from isothermal diagrams, and can be confirmed by metallographic examination. The results are then plotted on a temperature/cooling time diagram, which records, for example, the time to reach the beginning of the pearlite reaction over a range of cooling rates. This series of results will give rise to an austenite-pearlite boundary on the diagram and, likewise, lines showing the onset of the bainite transformation can be constructed. A schematic diagram is shown in Fig. 7.2 in which the boundaries for ferrite, pearlite, bainite and martensite are shown for a hypothetical steel. The diagram is best used by superimposing a transparent overlay sheet with the same scales and having lines representing various cooling rates drawn on it. The phases produced at a chosen cooling rate are then those which the superimposed line intersects on the continuous cooling diagram. In Fig. 7.2

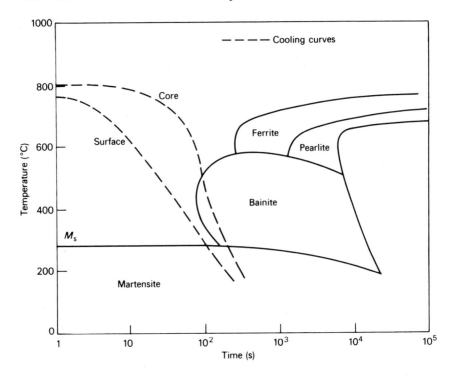

Fig. 7.2 Relation between cooling curves for the surface and core of an oil-quenched 95 mm diameter bar and the microstructure. The surface is fully martensitic (Thelning, *Steel and its Heat Treatment*, Bofors Handbook, Butterworths, 1975)

two typical cooling curves are superimposed for the surface and the centre of an oil-quenched 95 mm diameter bar. In this example, it should be noted that the centre cooling curve intersects the bainite region and consequently some bainite would be expected at the core of the bar after quenching in oil.

7.3 Hardenability testing

The rate at which austenite decomposes to form ferrite, pearlite and bainite is dependent on the composition of the steel, as well as on other factors such as the austenite grain size, and the degree of homogeneity in the distribution of the alloying elements. It is extremely difficult to predict hardenability entirely on basic principles, and reliance is placed on one of several practical tests, which allow the hardenability of any steel to be readily determined.

7.3.1 The Grossman test

Much of the earlier systematic work on hardenability was done by Grossman and coworkers who developed a test involving the quenching, in a particular cooling medium, of several cylindrical bars of different diameter of the steel under consideration. Transverse sections of the different bars on which hardness measurements have been made will show directly the effect of hardenability. In Fig. 7.3, which plots this hardness data for an SAE 3140 steel (1.1–1.4Ni, 0.55–0.75Cr, 0.40C) oil-quenched from 815°C, it is shown that the full martensitic hardness is only obtained in the smaller sections, while for larger diameter bars the hardness drops off markedly towards the centre of the bar. The softer and harder regions of the section can also be clearly resolved by etching.

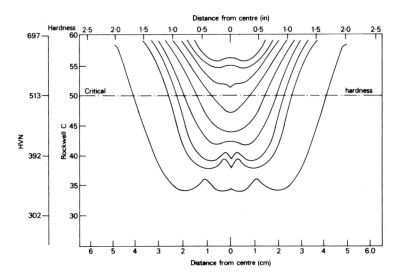

Fig. 7.3 1.1 Ni-0.75Cr-0.4C steel. Hardness data from transverse sections through water-quenched bars of increasing diameter (Grossman et al., In: Bain and Paxton, *Alloying Elements in Steel*, ASM 1961)

In the Grossman test, the transverse sections are metallographically examined to determine the particular bar which has 50% martensite at its centre. The diameter of this bar is then designated the critical diameter D_o. However, this dimension is of no absolute value in expressing the hardenability as it will obviously vary if the quenching medium is changed, e.g. from water to oil. It is therefore necessary to assess quantitatively the effectiveness of the different quenching media. This is done by determining coefficients for the severity of the quench usually referred to as H-coefficients. Typical values for three common quenching media and several conditions of agitation are shown in Table 7.1. The value for quenching in still water is set at 1, as a standard against which to compare other modes of quenching.

Using the H-coefficients, it is possible to determine in place of D_o, an ideal critical diameter D_i which has 50% martensite at the centre of the bar when

126 Steels—Microstructure and Properties

Table 7.1 H-coefficients of quenching media

Agitation	Cooling medium		
	Oil	Water	Brine
None	0.25–0.30	1.0	2.0
Moderate	0.35–0.40	1.2–1.3	
Violent	0.8–1.1	4.0	5.0

the surface is cooled at an infinitely rapid rate, i.e. when $H = \infty$. Obviously, in these circumstances $D_o = D_i$, thus providing the upper reference line in a series of graphs for different values of H (Fig. 7.4). In practice, H varies between about 0.2 and 5.0 (Table 7.1), so that if a quenching experiment is carried out at an H-value of, say, 0.4, and D_o is measured, then the graph of Fig. 7.4 can be used to determine D_i. This value will be a measure of the hardenability of a given steel, which is independent of the quenching medium used.

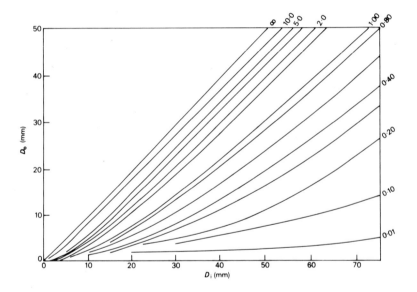

Fig. 7.4 Chart for determining ideal diameter (D_i) from the critical diameter (D_o) and the severity of quench (H) for carbon and medium alloy steels (Grossman and Bain, *Principles of Heat Treatment*, ASM, 1964)

7.3.2 The Jominy end quench test

While the Grossman approach to hardenability is very reliable, other less elaborate tests have been devised to provide hardenability data. Foremost amongst these is the Jominy test, in which a standardized round bar (25.4 mm diameter, 102 mm long) is heated to the austenitizing temperature, then placed on a rig in which one end of the rod is quenched by a standard jet of water (Fig. 7.5a, b). This results in a progressive decrease in the rate of cooling along the bar from the quenched end, the effects of which are determined by hardness measurements on flats ground 4 mm deep and parallel to the bar axis (Fig. 7.5c). A typical hardness plot for a steel containing 1% Cr 0.25% Mo 0.4% C (En 19B) is shown in Fig. 7.6, where the upper curve represents the hardness obtained with the upper limit of composition for the steel, while the lower curve is that for the composition at the lower limit. The area between the lines is referred to as a hardenability or Jominy band. Additional data, which is useful in conjunction with these results, is the hardness of quenched steels as a function both of carbon content and of the proportion of martensite in the structure. This data is given in Fig. 7.7 for as-quenched steels with 50–99% martensite. Therefore, the hardness for 50% martensite can be easily determined for a particular

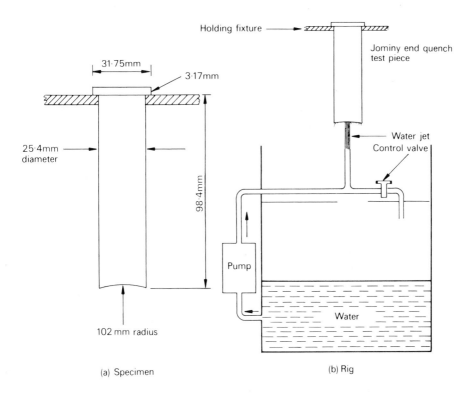

Fig. 7.5 The Jominy end quench test: a, specimen size; b, quenching rig (Wilson, *Metallurgy and Heat Treatment of Tool Steels*, McGraw Hill, 1975)

128 Steels—Microstructure and Properties

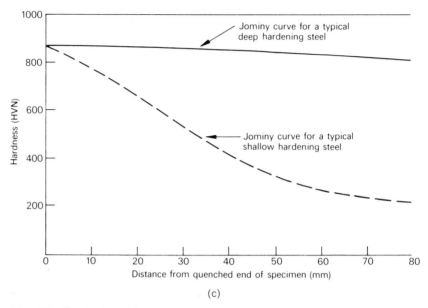

Fig. 7.5 The Jominy end quench test: c, Jominy hardness-distance curves for a shallow and a deep hardening steel (Wilson, *Metallurgy and Heat Treatment of Tool Steels*, McGraw Hill, 1975)

carbon content and, by inspection of the Jominy test results, the depth at which 50% martensite is achieved can be determined.

The Jominy test is now widely used to determine hardenabilities in the range $D_i = 1$–6; beyond this range the test is of limited use. The results can be readily converted to determine the largest diameter round bar which can be fully hardened. Fig. 7.8 plots bar diameter against the Jominy positions at which the same cooling rates as those in the centres of the bars are obtained for a series of different quenches. Taking the ideal quench ($H = \infty$), the highest curve, it can be seen that 12.5 mm along the Jominy bar gives a cooling rate equivalent to that at the centre of a 75 mm diameter bar. This diameter reduces to just over 50 mm for a quench in still water ($H = 1$). With, for example, a steel which gives 50% martensite at 19 mm from the quenched end after still oil quenching ($H = 0.3$), the critical diameter D_0 for a round rod will be 51 mm.

The diagram in Fig. 7.8 can also be used to determine the hardness at the centre of a round bar of a particular steel, provided a Jominy end quench test has been carried out. For example, if the hardness at the centre of a 5 cm diameter bar, quenched in still water, is required, Fig. 7.8 shows that this hardness will be achieved at about 12 mm along the Jominy test specimen from the quenched end. Reference to the Jominy hardness distance plot, then gives the required hardness value. If hardness values are required for

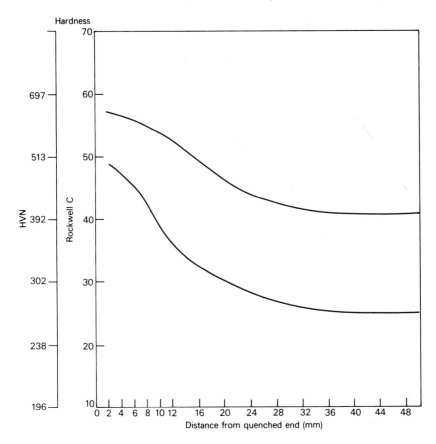

Fig. 7.6 Jominy curves for upper and lower limits of a steel, En 19B, giving a hardenability band (Thelning, *Steel and its Heat Treatment*, Bofors Handbook, Butterworths, 1975)

other points in round bars, e.g. surface or at half-radius, suitable diagrams are available for use.

7.4 Effect of grain size and chemical composition on hardenability

The two most important variables which influence hardenability are grain size and composition. The hardenability increases with increasing austenite grain size, because the grain boundary area is decreasing. This means that the sites for the nucleation of ferrite and pearlite are being reduced in number, with the result that these transformations are slowed down, and the hardenability is therefore increased. Likewise, most metallic alloying

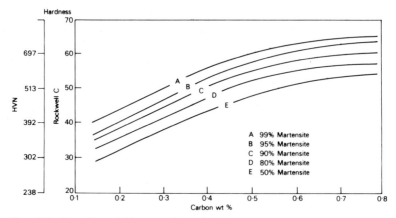

Fig. 7.7 The effect of % martensite and carbon content on as-quenched hardness (Hodge and Orohoski, In: Thelning, *Steel and its Heat Treatment*, Bofors Handbook, Butterworths, 1975)

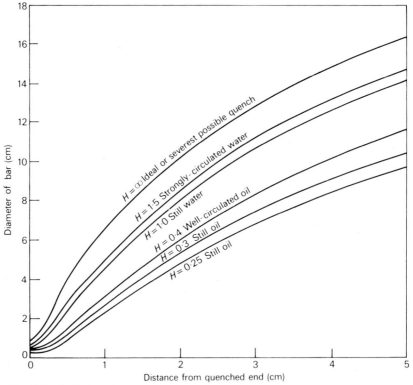

Fig. 7.8 Equivalent Jominy positions and bar diameter, where the cooling rate for the bar centre is the same as that for the point in the Jominy specimen. Curves are plotted for a range of cooling rates (Grossman and Bain, *Principles of Heat Treatment*, ASM, 1964)

elements slow down the ferrite and pearlite reactions, and so also increase hardenability. However, quantitative assessment of these effects is needed.

The first step is to determine the effect of grain size and of carbon content. Data is available, in so far as D_i has been determined for steels with carbon in the range 0.2–1%, and for a range of grain sizes (ASTM 4–8), as shown in Fig. 7.9. Use of this diagram for any steel provides a base hardenability figure, D_{iC}, which must be modified by taking into account the effect of additional alloying elements. This is done by use of multiplying factors

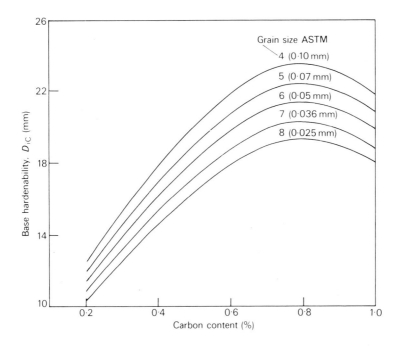

Fig. 7.9 Effect of carbon content and grain size on base hardenability (Moser and Legat, *Härterei Techn. Mitt.*, 1969, **24**, 100)

which have been experimentally determined for the familiar alloying elements (Fig. 7.10). The ideal critical diameter D_i is then found from the empirical relationship:

$$D_i = D_{iC} \times 2.21(\%\text{Mn}) \times 1.40(\%\text{Si}) \times 2.13(\%\text{Cr}) \times 3.275(\%\text{Mo}) \\ \times 1.47(\%\text{Ni}) \text{ (weight percentages)} \qquad (7.1)$$

This relationship, due to Moser and Legat, appears to be more accurate in practice than a much earlier one put forward by Grossman. Further corrections have to be made for different austenitizing temperatures when dealing with high carbon steels, but, on the whole, the relationship is quite effective in predicting actual hardening behaviour.

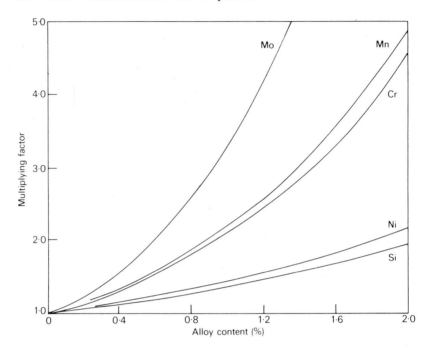

Fig. 7.10 Hardenability multiplying factors for common alloying elements (Moser and Legat, *Härterei Techn. Mitt.*, 1969, **24**, 100)

It is also possible to calculate Jominy hardness/distance curves from the chemical composition using regression analysis, but the Jominy test is relatively straightforward to carry out, and is often used as a quality control technique. Therefore, there is probably not a great need for an empirical relationship to serve the purpose, even if all the effective variables could be properly accounted for.

7.5 Jominy tests and continuous cooling diagrams

Since a continuous cooling diagram shows the phases which will occur in a particular steel over a range of different cooling rates, it should be possible to correlate with such diagrams the information obtained from a Jominy test. For example, we know the rates of cooling characteristic of each point on a Jominy specimen, so cooling curves corresponding to selected points can be superimposed on a CCT diagram. Consequently we can identify the phases to be expected at these points along a Jominy test specimen. By relating directly the cooling rates to those existing at specific points in

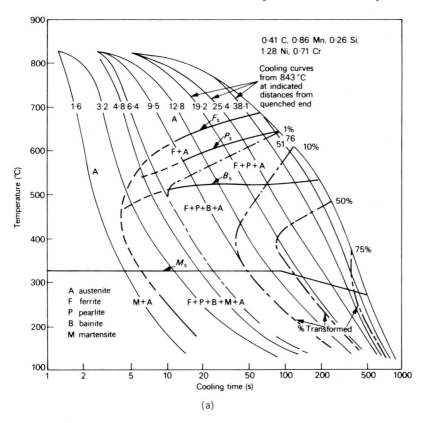

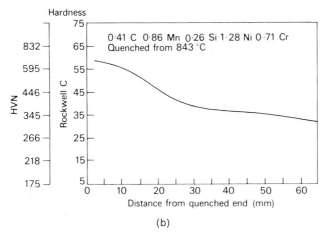

Fig. 7.11 a, Continuous cooling transformation (CCT) diagram for a steel with 1.28Ni, 0.7Cr, 0.86Mn, 0.26Si, 0.41C; b, Jominy hardness for the same steel (Bethlehem Steel Co., *Metal Progress*, October 1963)

normally quenched bars (Fig. 7.8) the structure can be predicted at these points. Likewise, from the Jominy hardness data, the hardness of the particular structure can also be specified. A typical CCT diagram for a NiCr steel is represented in Fig. 7.11a together with the Jominy hardness for the same material (Fig. 7.11b), which shows that this steel will only be fully martensitic for the first 2 mm of the Jominy specimen. Beyond this distance, ferrite, pearlite and bainite begin to appear in the microstructure.

Strictly speaking, the cooling rates at particular points on a Jominy specimen will vary to some extent as the steel composition is altered because changes occur in thermal conductivity, but for low alloy steels the correction is not a large one.

7.6 Hardenability and heat treatment

There is a bewildering number of steels, the compositions of which are usually complex and defined in most cases by specifications which give ranges of concentration of the important alloying elements, together with the upper limits of impurity elements such as sulphur and phosphorus. It is not the purpose of this book to examine how a particular steel is selected for a given application, except to say that it is usually a result of cumulative experience and of an evolutionary approach which accepts new steels after thorough testing and practical trial over a long period of time.

While alloying elements are used for various reasons, the most important is the achievement of higher strength in required shapes and sizes and often in very large sections which may be up to a metre or more in diameter in the case of large shafts and rotors. Hardenability is, therefore, of the greatest importance, and one must aim for the appropriate concentrations of alloying element needed to harden fully the section of steel under consideration. Equally, there is little point in using too high a concentration of alloying element, i.e. more than that necessary for full hardening of the required sections. Alloying elements are usually much more expensive than iron, and in some cases are a diminishing natural resource, so there is additional reason to use them effectively in heat treatment. Carbon has a marked influence on hardenability, but its use at higher levels is limited, because of the lack of toughness which results, as well as the greater difficulties in fabrication and, most important, increased probability of distortion and cracking during heat treatment and welding.

The most economical way of increasing the hardenability of a plain carbon steel is to increase the manganese content, an increase from 0.60 wt% to 1.40 wt%, giving a substantial improvement in hardenability. Chromium and molybdenum are also very effective, and amongst the cheaper alloying additions per unit of increased hardenability. Boron has a particularly large effect when added to a fully deoxidized low carbon steel, even in concentrations of the order of 0.001%, and would be more widely used if its distribution in steel could be more easily controlled. The role of grain size should not be overlooked because an increase in grain diameter from 0.02 mm to 0.125 mm can increase the hardenability by as much as 50%, which is very acceptable provided the mechanical properties, particularly toughness, are not too adversely affected.

Hardenability data now exists for a wide range of steels in the form of maximum and minimum end-quench hardenability curves, usually referred to as hardenability bands. This data is available for very many of the steels listed in specifications such as those of the American Society of Automotive Engineers (SAE), the American Iron and Steel Institute (AISI) and the British Standard En and DTD series of steels. Fig. 7.12 gives some typical data for 0.5Cr 0.5Ni 0.25Mo type steels for the carbon range, 0.20–0.60%. The curves in the figure are minimum hardenability limits.

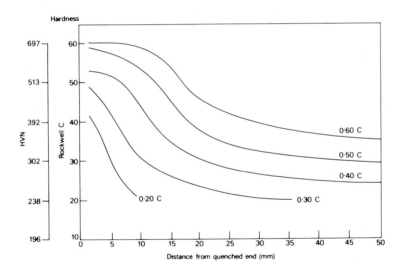

Fig. 7.12 Effect of increasing carbon on the hardenability of a 0.5Cr-0.5Ni-0.25Mo steel: minimum hardenability limits (US Steel Co., *The Making, Shaping and Treating of Steel*, 1971)

High hardenability is not always desirable for many tool and machine parts, where a hard wear resistant surface is best combined with a tough core. Such shallow hardening situations are additionally preferred because, on quenching, the core develops a tensile internal stress while the surface becomes stressed in compression. This situation is very desirable because any fatigue cracks nucleated at surface stress concentrations will find propagation more difficult when a compressive stress is present.

7.7 Quenching stresses and quench cracking

The act of quenching from the austenitic region to room temperature is a drastic treatment which all too often leads to distortion in the part and even serious cracking (*quench cracking*). These defects arise from internal stresses which develop during quenching from two sources:

(1) *thermal stresses* arising directly from the different cooling rates experienced by the surface and the interior of the steel

(2) *transformation stresses* due to the volume changes which occur when austenite transforms to other phases.

An example of the effect of thermal stresses is given in Fig. 7.13 for a steel bar 10 cm in diameter quenched into water from 850°C. The temperature-time relationship for the surface and the core are given in Fig. 7.13a, from which it is seen that the maximum temperature difference occurs after a time t, when it is about 500°C, which could give rise to a stress in excess of 1000 MN m^{-2}, if no relaxation took place. Under these conditions, the surface stress-time relationship would be that of curve A, Fig. 7.13b. However, the maximum stress level is not sustained because plastic deformation takes place and the stress-time relationship in reality is that indicated by curve B. The tensile stress in the surface is balanced by a compressive stress in the core as shown by curve C. At some lower

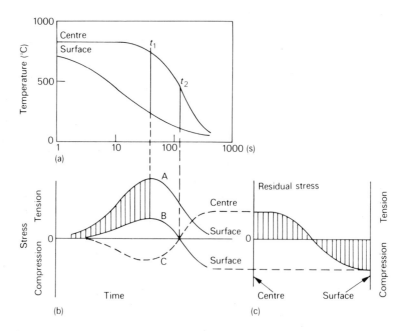

Fig. 7.13 Development of thermal stresses during cooling of a 10 cm diameter bar quenched into water from 850°C (after Rose, *Härterei Techn. Mitt.*, 1966, **21**, 1)

temperature t_2 the compressive and tensile stresses will both fall to zero but as the temperature drops further to room temperature the stress situation reverses and the core goes into tension and the surface into compression. Fig. 7.13c shows the stress distribution through the bar at room temperature.

The more rapid the quench, the higher the temperature difference between core and surface during quenching and, therefore, the higher the resulting stresses at room temperature. In practical terms this means that avoidance of distortion involves the use of less drastic quenching media, e.g. oil instead of water, and consequently adjustments have to be made to the hardenability if full hardening through the section is required.

Transformation stresses arise from the change in volume associated with the formation of a new phase. For example, when austenite transforms to martensite in a 1% carbon steel, there is an increase in volume of 4%, while the transformation to pearlite results in an increase of 2.4%. The effect of these volume changes on the stress pattern developed depends on whether the reaction at surface and core start simultaneously, and whether the hardenability is sufficient to permit full hardening or not. If the martensite reaction starts at the surface, a tensile stress is generated there and a compressive stress occurs at the centre, a situation which is accentuated by having the martensite reaction throughout the diameter, i.e. in small sections, or in steels of high hardenability. The presence of a tensile stress in the surface is not advisable for reasons given above, so it is clear that in some cases high hardenability can create problems. These can be avoided by the use of steels which provide only a relatively thin hardened layer at the surface which can be maintained in a state of compression. Surface treatment methods such as carburizing and nitriding, where the interstitial element concentration is substantially increased by a diffusive process, not only lead to hard wear resistant surfaces, but also surfaces which resist crack propagation by being subject to compressive stresses.

Martensite is a very brittle phase which becomes more brittle with increasing carbon content. In higher carbon martensites, which tend to exhibit the burst phenomenon in which individual martensite plates are successively nucleated by previous plates, cracks are often observed in plates at points of impact of later plates upon them. These micro-cracks provide obvious nuclei for the propagation of major cracks. In broader terms, quench cracking is likely to occur when quenching stresses have not been sufficiently released by plastic deformation at elevated temperatures, and they therefore reach the fracture stress of the steel. As in the case of fatigue cracking, the safest situation is to have the most sensitive region of the steel under compressive stresses.

There are some fairly obvious precautions which can be taken to avoid such cracking, including the use of the slowest quench compatible with the achievement of adequate hardenability. Also stress concentrations in the form of notches, heavy machining grooves and sudden changes in cross section should be avoided where possible, as these will all encourage quench-crack nucleation.

The composition of the steel is important because the transformation characteristics will influence the incidence of cracking. The effect of carbon has already been referred to but, additionally, the M_s temperature decreases with increasing carbon content. Thus, in higher carbon steels, the quenching stresses are less likely to be relieved than would be the case if the martensite begins to form at a higher temperature where the steel is more able to relieve stresses by flow than by fracture. Further, the lower the M_s temperature the larger the change in volume during the transformation and, therefore, the higher the transformation stresses developed. Metallic alloying elements also depress the M_s, but by substantially increasing the hardenability they allow the use of less drastic quenching which greatly reduces the probability of distortion and cracking.

7.8 Martempering

A very effective way of reducing the quenching stresses is to interrupt the quench just above the M_s in the metastable austenite region of the TTT diagram (Fig. 7.14), usually by quenching into a salt bath held at the

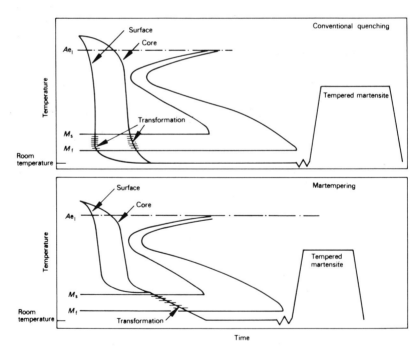

Fig. 7.14 TTT curves and superimposed cooling curves comparing conventional quenching with martempering. (*Metals Handbook*, 8th edition, Volume 2, ASM)

appropriate temperature. When the part has uniformly reached the temperature of the bath, and before any isothermal transformation has taken place, the steel is cooled in air through the martensitic range to room temperature. This interrrupted quenching process is referred to as *martempering*, and sometimes as mar-quenching. However, it does not involve tempering and the martensite has to be tempered in the normal way. This process can be very successful in minimizing distortion but requires, for its successful application, a steel of adequate hardenability so that the pearlitic and bainitic reactions can be avoided. An alloy steel is usually used which would be normally hardened by quenching in oil. However, for effective use, it is essential to know accurately the temperature range of martensite formation and the effect of austenite grain size on this range.

Further reading

Holloman, J. H. and Jaffe, L. D., *Ferrous Metallurgical Design*, Wiley, 1947

Bullens, D. K. and Battelle Staff, *Steel and its Heat Treatment*, Volumes I–II, Wiley, 5th edition, 1948; Volume III, 1949.

United States Steel Corporation, *The Making, Shaping and Treating of Steel*, 9th edition, 1971.

Bain, E. C. and Paxton, H. W., *Alloying Elements in Steel*, American Society for Metals, 2nd edition, 1961

Grossman, M. A. and Bain, E. C., *Principles of Heat Treatment*, American Society for Metals, 5th edition, 1964

American Society for Metals, *Quenching and Martempering*, 1964

Rose, A., Internal stresses resulting from heat treatment and transformation processes, *Härterei Techn. Mitt.*, **21**, (1), 1, 1966

Transformation and Hardenability in Steel, Symposium published by Climax Molybdenum, 1968

Thelning, K. E., *Steel and its Heat Treatment*, Bofors Handbook, Butterworth, 1975

Wilson, Robert, *Metallurgy and Heat Treatment of Tool Steels*, McGraw Hill, 1975

Doane, D. V. and Kirkaldy, J. S. (eds), *Hardenability Concepts with Applications to Steel*, The Metallurgical Society of AIME, 1978

8
The tempering of martensite

8.1 Introduction

Martensite is a very strong phase but it is normally very brittle so it is necessary to modify the mechanical properties by heat treatment in the range 150–700°C. This process, which is called *tempering*, is one of the oldest heat treatments applied to steels although it is only in recent years that a detailed understanding of the phenomena involved has been reached. Essentially, martensite is a highly supersaturated solid solution of carbon in iron which, during tempering, rejects carbon in the form of finely divided carbide phases. The end result of tempering is a fine dispersion of carbides in an α-iron matrix which often bears little structural similarity to the original as-quenched martensite.

It should be borne in mind that, in many steels, the martensite reaction does not go to completion on quenching, resulting in varying amounts of retained austenite which does not remain stable during the tempering process.

8.2 Tempering of plain carbon steels

The as-quenched martensite possesses a complex structure which has been referred to in Chapter 5. The laths or plates are heavily dislocated to an extent that individual dislocations are very difficult to observe in thin-foil electron micrographs. A typical dislocation density for a 0.2 carbon steel is between 0.3 and 1.0×10^{12} cm cm^{-3}. As the carbon content rises above about 0.3%, fine twins about 5–10 nm wide are also observed. Often carbide particles, usually rods or small plates, are observed (Fig. 8.1). These occur in the first-formed martensite, i.e. the martensite formed near M_s, which has the opportunity of tempering during the remainder of the quench. This phenomenon, which is referred to as *auto-tempering*, is clearly more likely to occur in steels with a high M_s.

On reheating as-quenched martensite, the tempering takes place in four distinct but overlapping stages:

Stage 1, up to 250°C–precipitation of ε-iron carbide; partial loss of tetragonality in martensite
Stage 2, between 200 and 300°C–decomposition of retained austenite

The tempering of martensite 141

Fig. 8.1 Fe-0.2C quenched from 1100°C into iced brine. Autotempered martensite (Ohmori). Thin-foil EM

Stage 3, between 200 and 350°C–replacement of ε-iron carbide by cementite; martensite loses tetragonality

Stage 4, above 350°C–cementite coarsens and spheroidizes; recrystallization of ferrite.

8.2.1 Tempering–stage 1

Martensite formed in medium and high carbon steels (0.3–1.5%C) is not stable at room temperature because interstitial carbon atoms can diffuse in the tetragonal martensite lattice at this temperature. This instability increases between room temperature and 250°C, when ε-iron carbide precipitates in the martensite (Fig. 8.2). This carbide has a close-packed hexagonal structure, and precipitates as narrow laths or rodlets on cube planes of the matrix with a well-defined orientation relationship (Jack):

$(101)_{\alpha'}//(10\bar{1}1)_{\varepsilon}$
$(011)_{\alpha'}//(0001)_{\varepsilon}$
$[11\bar{1}]_{\alpha'}//[12\bar{1}0]_{\varepsilon}$

X-ray measurements indicate that the lattice spacings of $(101)_{\alpha}$ and $(10\bar{1}1)_{\varepsilon}$ are within about 0.5%, so lattice coherency is likely in the early stages of precipitation. In fact, in the higher carbon steels, an increase in hardness has been observed on tempering in the range 50–100°C, which is attributed to precipitation hardening of the martensite by ε-carbide. At the

142 *Steels—Microstructure and Properties*

Fig. 8.2 Fe-0.8C quenched and tempered at 250°C. Precipitates of ε carbide and cementite (arrowed) (Ohmori). Thin-foil EM

end of stage 1 the martensite still possesses a tetragonality, indicating a carbon content of around 0.25%. It follows that steels with lower carbon contents are unlikely to precipitate ε-carbide. This stage of tempering possess an activation energy of between 60 and 80 kJ mol^{-1}, which is in the right range for diffusion of carbon in martensite. The activation energy has been shown to increase linearly with the carbon concentration between 0.2 and 1.5 wt%C. This would be expected as increasing the carbon concentration also increases the occupancy of the preferred interstitial sites, i.e. the octahedral interstices at the mid-points of unit cell edges, and centres of cell faces, thus reducing the mobility of the C atoms.

8.2.2 Tempering–stage 2

During stage 2, austenite retained during quenching is decomposed, usually in the temperature range 230–300°C. Cohen and coworkers were able to detect this stage by X-ray diffraction measurements as well as dilatometric and specific volume measurements. However, the direct observation of retained austenite in the microstructure has always been rather difficult, particularly if it is present in low concentrations. In martensitic plain carbon steels below 0.5% carbon, the retained austenite is often below 2%, rising to around 6% at 0.8%C and over 30% at 1.25%C. The little available evidence suggests that in the range 230–300°C, retained austenite decomposes to bainitic ferrite and cementite, but no detailed comparison between this phase and lower bainite has yet been made.

8.2.3 Tempering–stage 3

During the third stage of tempering, cementite first appears in the microstructure as a Widmanstätten distribution of rods which have a well-defined orientation relationship with the matrix which has now lost its tetragonality and become ferrite. The relationship is that due to Bagaryatski:

$(211)_{\alpha'} // (001)_{Fe_3C}$
$[01\bar{1}]_{\alpha'} // [100]_{Fe_3C}$
$[\bar{1}11]_{\alpha'} // [010]_{Fe_3C}$

This reaction commences as low as 100°C, and is fully developed at 300°C, with particles up to 200 nm long and ~15 nm in diameter. Similar structures are often observed in lower carbon steels as quenched, as a result of the formation of Fe_3C during the quench. During tempering, the most likely sites for the nucleation of the cementite are the ε-iron carbide interfaces with the matrix (Fig. 8.2), and as the Fe_3C particles grow, the ε-iron carbide particles gradually disappear.

The twins occurring in the higher carbon martensites are also sites for the nucleation and growth of cementite which tends to grow along the twin boundaries forming colonies of similarly oriented lath-shaped particles (Fig. 8.3) of $\{112\}_\alpha$ habit, which can be readily distinguished from the

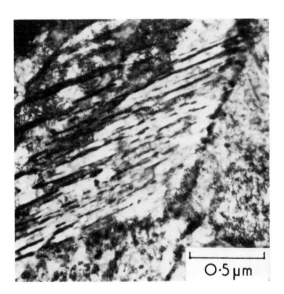

Fig. 8.3 Fe-0.8C quenched and tempered at 450°C. Fe₃C growing along twin boundaries (Ohmori). Thin-foil EM

normal Widmanstätten habit. The orientation relationship with the ferritic matrix is the same in both these cases.

A third site for the nucleation of cementite is the grain boundary regions (Fig. 8.4), both the interlath boundaries of the martensite and the original austenite grain boundaries. The cementite can form as very thin films which are difficult to detect but which gradually spheroidize to give rise to well-defined particles of Fe_3C in the grain boundary regions. There is some evidence to show that these grain boundary cementite films can adversely affect ductility. However, they can be modified by addition of alloying elements.

Fig. 8.4 Fe-0.8C quenched and tempered at 250°C. Grain boundary precipitation of cementite (Ohmori). Thin-foil EM

During the third stage of tempering the tetragonality of the matrix disappears and it is then, essentially, ferrite, not supersaturated with respect to carbon. Subsequent changes in the morphology of the cementite particles occur by an Ostwald ripening type of process, where the smaller particles dissolve in the matrix providing carbon for the selective growth of the larger particles.

8.2.4 Tempering–stage 4

It is useful to define a fourth stage of tempering in which the cementite particles undergo a coarsening process and essentially lose their crystallographic morphology, becoming spheroidized. The coarsening commences between 300 and 400°C, while spheroidization takes place increasingly up to 700°C. At the higher end of this range of temperature the martensite lath boundaries are replaced by more equi-axed ferrite grain boundaries by a process which is best described as recrystallization. The final result is an equi-axed array of ferrite grains with coarse spheroidized particles of Fe_3C (Fig. 8.5), partly, but not exclusively, in the grain boundaries.

Fig. 8.5 Fe-0.17C water quenched from 900°C and tempered 5h at 650°C. Spheroidized Fe_3C in equi-axed ferrite (lenel). Optical micrograph, ×350

The spheroidization of the Fe_3C rods is encouraged by the resulting decrease in surface energy. The particles which preferentially grow and spheroidize are located mainly at interlath boundaries and prior austenite boundaries, although some particles remain in the matrix. The boundary sites are preferred because of the greater ease of diffusion in these regions. Also, the growth of cementite into ferrite is associated with a decrease in density so vacancies are required to accommodate the growing cementite. Vacancies will diffuse away from cementite particles which are redissolving in the ferrite and towards cementite particles which are growing, so that the rate-controlling process is likely to be the diffusion of vacancies. The measured activation energies are much higher (210–315 kJ mol^{-1}), than that for diffusion of carbon in ferrite (~ 84 kJ mol^{-1}), and much closer to the activation energy for self diffusion in α-iron (~ 250 kJ mol^{-1}).

The original martensite lath boundaries remain stable up to about 600°C, but in the range 350–600°C, there is considerable rearrangement of the dislocations within the laths and at those lath boundaries which are essentially low angle boundaries. This leads to a marked reduction in the dislocation density and to lath-shaped ferritic grains closely related to the packets of similarly oriented laths in the original martensite. This process, which is essentially one of recovery, is replaced between 600 and 700°C by recrystallization which results in the formation of equi-axed ferrite grains with spheroidal Fe_3C particles in the boundaries and within the grains. This process occurs most readily in low carbon steels. At higher carbon contents the increased density of Fe_3C particles is much more effective in pinning the ferrite boundaries, so recrystallization is much more sluggish. The final process is the continued coarsening of the Fe_3C particles and gradual ferrite grain growth (Fig. 8.6).

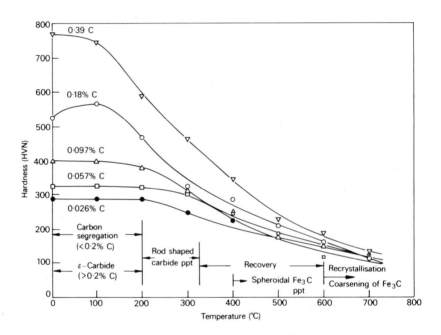

Fig. 8.6 Hardness of iron-carbon martensites tempered 1h at 100–700°C (Speich, *Trans. Met. Soc. AIME*, 1969, **245**, 2553).

8.2.5 Role of carbon content

Carbon has a profound effect on the behaviour of steels during tempering. Firstly, the hardness of the as-quenched martensite is largely influenced by the carbon content (Fig. 8.6), as is the morphology of the martensite laths which have a $\{111\}_\gamma$ habit plane up to 0.3%C, changing to $\{225\}_\gamma$ at higher carbon contents. The M_s temperature is reduced as the carbon content increases, and thus the probability of the occurrence of auto-tempering is less. During fast quenching in alloys with less than 0.2%C, the majority (up to 90%) of the carbon segregates to dislocations and lath boundaries, but with slower quenching some precipitation of cementite occurs. On subsequent tempering of low carbon steels up to 200°C further segregation of carbon takes place, but no precipitation has been observed. Under normal circumstances it is difficult to detect any tetragonality in the martensite in steels with less than 0.2%C, a fact which can also be explained by the rapid segregation of carbon during quenching.

The hardness changes during tempering are also very dependent on carbon content, as shown in Fig. 8.6 for steels up to 0.4%C. Above this concentration, an increase in hardness has been observed in the temperature range 50–150°C, as ε-carbide precipitation strengthens the martensite. However, the general trend is an overall softening, as the tempering temperature is raised. The diagram indicates the main physical processes contributing to the change in mechanical properties.

8.3 Mechanical properties of tempered plain carbon steels

The intrinsic mechanical properties of tempered plain carbon martensitic steels are difficult to measure for several reasons. Firstly, the absence of other alloying elements means that the hardenability of the steels is low, so a fully martensitic structure is only possible in thin sections. However, this may not be a disadvantage where shallow hardened surface layers are all that is required. Secondly, at lower carbon levels, the M_s temperature is rather high, so autotempering is likely to take place. Thirdly, at the higher carbon levels the presence of retained austenite will influence the results. Added to these factors, plain carbon steels can exhibit quench cracking which makes it difficult to obtain reliable test results. This is particularly the case at higher carbon levels, i.e. above 0.5% carbon.

Provided care is taken, very good mechanical properties, in particular proof and tensile stresses, can be obtained on tempering in the range 100–300°C. However, the elongation is frequently low and the impact values poor. Table 8.1 shows some typical results for plain carbon steels with between 0.2 and 0.5%C, tempered at low temperatures.

Plain carbon steels with less than 0.25%C are not normally quenched and tempered, but in the range 0.25–0.55%C heat treatment is often used to upgrade mechanical properties. The usual tempering temperature is between 300 and 600°C allowing the development of tensile strengths between 1700 and 800 MN m^{-2}, the toughness increasing as the tensile strength decreases. This group of steels is very versatile as they can be used for crankshafts and general machine parts as well as hand tools, such as screwdrivers and pliers.

The high carbon steels (0.5–1.0%) are much more difficult to fabricate

Table 8.1 Mechanical properties of plain carbon steels, both as-quenched and tempered (after Irvine, K. J. et al., J.I.S.I., 1960, **196**, 70)

Steel	Property	Treatment			
		As-quenched	Tempered 7 hours at		
			100°C	200°C	300°C
0.2%C	0.2% Proof stress (MNm^{-2})	1270	1460	1235	1110
0.3%C		1360	1370	1270	1140
0.5%C				1670	1410
0.2%C	UTS (MNm^{-2})	1470	1690	1450	1340
0.3%C		1580	1605	1460	1240
0.5%C				2040	1600
0.2%C	Elongation (%)	5.0	6.0	6.0	9.0
0.3%C		4.5	7.0	7.0	10.0
0.5%C				4.0	7.0
0.2%C	Hardness (DPN)	446	444	446	357
0.3%C		564	517	502	420
0.5%C		680	666	571	470

and are, therefore, particularly used in applications where high hardness and wear resistance are required, e.g. axes, knives, hammers, cutting tools. Typical mechanical properties as a function of tempering temperature are shown in Fig. 8.7 for a steel at the lower level (0.5%C) of this range. Another important application is for springs, where often the required mechanical properties are obtained simply by heavy cold work, i.e. hard drawn spring wire. However, carbon steels in the range 0.5–0.75%C are quenched, then tempered to the required yield stress.

8.4 Tempering of alloy steels

The addition of alloying elements to a steel has a substantial effect on the kinetics of the $\gamma \rightarrow \alpha$ transformation, and also of the pearlite reaction. Most common alloying elements move the TTT curves to longer times, with the result that it is much easier to 'miss' the nose of the curve during quenching. This essentially gives higher hardenability, since martensite structures can be achieved at slower cooling rates and, in practical terms, thicker specimens can be made fully martensitic. Alloying elements have also been shown to have a substantial effect in depressing the M_s temperature. In this section, we will examine the further important effects of alloying elements during the tempering of martensite, where not only the kinetics of the basic reactions are influenced but also the products of these reactions can be substantially changed, e.g. cementite can be replaced by other carbide

The tempering of martensite   149

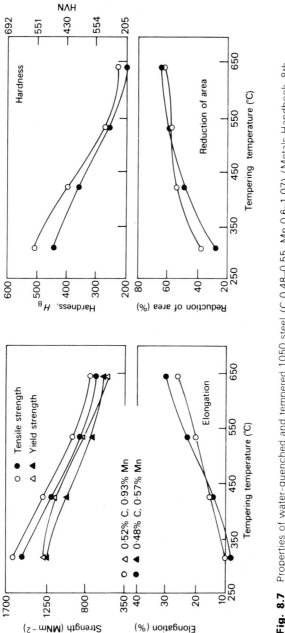

Fig. 8.7 Properties of water-quenched and tempered 1050 steel (C 0.48–0.55, Mn 0.6–1.07) (Metals Handbook, 8th edition, Volume 1, ASM)

phases. Several of the simpler groups of alloy steels will be used to provide examples of the general behaviour.

8.4.1 The effect of alloying elements on the formation of iron carbides

The structural changes during the early stage of tempering are difficult to follow. However, it is clear that certain elements, notably silicon, can stabilize the ε-iron carbide to such an extent that it is still present in the microstructure after tempering at 400°C in steels with 1–2% Si, and at even higher temperatures if the silicon is further increased. The evidence suggests that both the nucleation and growth of the carbide is slowed down and that silicon enters into the ε-carbide structure. It is also clear that the transformation of ε-iron carbide to cementite is delayed considerably. While the tetragonality of martensite disappears by 300°C in plain carbon steels, in steels containing some alloying elements, e.g. Cr, Mo, W, V, Ti, Si, the tetragonal lattice is still observed after tempering at 450°C and even as high as 500°C. It is clear that these alloying elements increase the stability of the supersaturated iron-carbon solid solution. In contrast manganese and nickel decrease the stability (Fig. 8.8).

Alloying elements also greatly influence the proportion of austenite retained on quenching. Typically, a steel with 4% molybdenum, 0.2%C, in the martensitic state contains less than 2% austenite, and about 5% is detected in a steel with 1% vanadium and 0.2%C. The austenite can be revealed as a fine network around the martensite laths, by using dark field electron microscopy (Fig. 8.9). On tempering each of the above steels at 300°C, the austenite decomposes to give thin grain boundary films of cementite which, in the case of the higher concentrations of retained austenite, can be fairly continuous along the lath boundaries. It is likely that this interlath cementite is responsible for *tempered martensite embrittlement*, frequently encountered as a toughness minimum in the range 300–350°C, by leading to easy nucleation of cracks, which then propagate across the tempered martensite laths.

Alloying elements can also restrain the coarsening of cementite in the range 400–700°C, a basic process during the fourth stage of tempering. Several alloying elements, notably silicon, chromium, molybdenum and tungsten, cause the cementite to retain its fine Widmanstätten structure to higher temperatures, either by entering into the cementite structure or by segregating at the carbide-ferrite interfaces. Whatever the basic cause may be, the effect is to delay significantly the softening process during tempering. This influence on the cementite dispersion has other effects, in so far as the carbide particles, by remaining finer, slow down the reorganization of the dislocations inherited from the martensite, with the result that the dislocation substructures refine more slowly. The cementite particles are also found on ferrite grain boundaries, where they control the rate at which the ferrite grains grow. Gladman has shown, for a precipitate of volume

The tempering of martensite 151

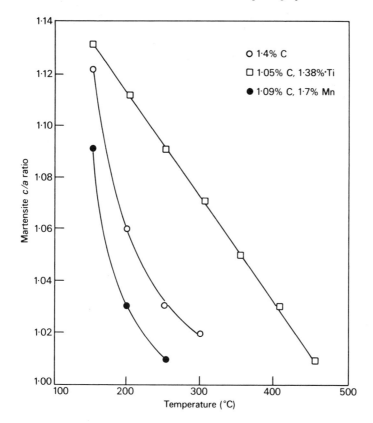

Fig. 8.8 Effect of Ti and Mn on the tetragonality of martensite during tempering (Kurdjumov, *JISI*, 1960, **195**, 26)

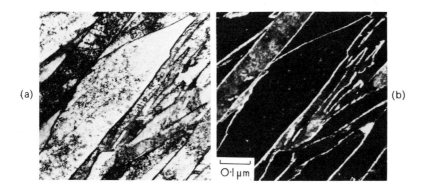

Fig. 8.9 Fe-10Cr-0.2C quenched in iced brine from 1150°C. Martensite with retained austenite (Howell): a, bright field EM; b, dark field EM using γ-diffraction beam. The γ areas are light

fraction f, pinning polygonal grains of average radius r_o, that the critical radius of particle r_{crit}, before grain growth can occur is

$$r_{crit} = \frac{6r_o f}{\pi}\left(\frac{3}{2} - \frac{2}{Z}\right)^{-1} \tag{8.1}$$

where Z is the ratio of radii of matrix and growing grains. This expression has been found to fit well the experimental results on silicon steels.

In plain carbon steels cementite particles begin to coarsen in the temperature range 350–400°C, and addition of chromium, silicon, molybdenum or tungsten delays the coarsening to the range 500–550°C. It should be emphasized that up to 500°C, the only carbides to form are those of iron. However, they will take varying amounts of alloying elements into solid solution and may reject other alloying elements as they grow.

8.4.2 The formation of alloy carbides–secondary hardening

A number of the familiar alloying elements in steels form carbides which are thermodynamically more stable than cementite. It is interesting to note that this is also true of a number of nitrides and borides. Nitrogen and boron are increasingly used in steels in small but significant concentrations. The enthalpies of formation of some of these compounds are shown in Fig. 4.5 (p. 60), in which iron carbide is the least stable compound situated at the right of the diagram. The alloying elements Cr, Mo, V, W and Ti all form carbides with substantially higher enthalpies of formation, while the elements nickel, cobalt and copper do not form carbide phases. Manganese is a weak carbide former, found in solid solution in cementite and not in a separate carbide phase.

It would, therefore, be expected that when strong carbide forming elements are present in a steel in sufficient concentration, their carbides would be formed in preference to cementite. Nevertheless, during the tempering of all alloy steels, alloy carbides do not form until the temperature range 500–600°C, because below this the metallic alloying elements cannot diffuse sufficiently rapidly to allow alloy carbides to nucleate. The metallic elements diffuse substitutionally, in contrast to carbon and nitrogen which move through the iron lattice interstitially, with the result that the diffusivities of carbon and nitrogen are several orders of magnitude greater in iron, than those of the metallic alloying elements (Table 1.4). Consequently, higher temperatures are needed for the necessary diffusion of the alloying elements prior to the nucleation and growth of the alloy carbides and, in practice, for most of the carbide forming elements this is in the range 500–600°C.

The coarsening of carbides in steels is an important phenomenon which influences markedly the mechanical properties. We can apply in general terms the theory for coarsening of a dispersion due to Lifshitz and Wagner, which gives for spherical particles in a matrix:

$$r_t^3 - r_o^3 = \frac{k}{RT} V_m^2 D\gamma t \qquad (8.2)$$

where r_o = the mean particle radius at time zero
 r_t = the mean particle radius at time t
 D = diffusion coefficient of solute in matrix
 γ = interfacial energy of particle/matrix interface
 V_m = molar volume of precipitate
 k = constant.

The coarsening rate is dependent on the diffusion coefficient of the solute and, under the same conditions, at a given temperature, cementite would coarsen at a greater rate than any of the alloy carbides once formed (see Section 8.2.4 where the role of vacancies is discussed). This occurs in alloy steels in which cementite and an alloy carbide coexist, where the cementite dispersion is always much coarser. It is this ability of certain alloying elements to form fine alloy carbide dispersions in the range 500–600°C, which remain very fine even after prolonged tempering, that allows the development of high strength levels in many alloy steels. Indeed, the formation of alloy carbides between 500 and 600°C is accompanied by a marked increase in strength, often in excess of that of the as-quenched martensite (Fig. 8.10). This phenomenon, which is referred to as *secondary hardening*, is best shown in steels containing molybdenum, vanadium, tungsten, titanium, and also in chromium steels at higher alloy concentrations.

This secondary hardening process is a type of age-hardening reaction, in which a relatively coarse cementite dispersion is replaced by a new and much finer alloy carbide dispersion. On attaining a critical dispersion parameter, the strength of the steel reaches a maximum, and as the carbide dispersion slowly coarsens, the strength drops. The process is both time and temperature dependent, so both variables are often combined in a parameter

$$P = T(k + \log t) \qquad (8.3)$$

where T is the absolute temperature and t the tempering time, while k is a constant which is about 20 for alloy steels. Usually referred to as the Holloman-Jaffe parameter, this can be plotted against hardness to give one typical curve for a particular steel. In Fig. 8.10 the effect of increasing molybdenum content is thus effectively demonstrated in a series of steels containing 0.1% carbon. Significantly, noncarbide-forming elements such as nickel, cobalt, silicon, do not give secondary hardening. However, some elements, e.g. silicon, by delaying the coarsening of cementite, lead to a plateau on the tempering curve in the range 300–500°C.

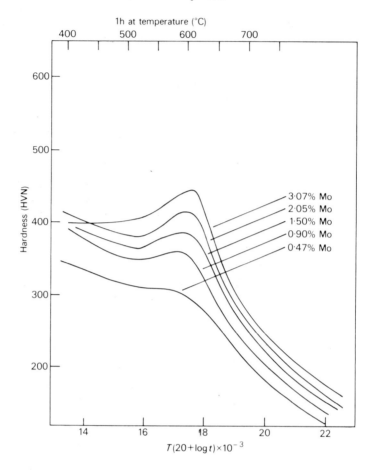

Fig. 8.10 The effect of molybdenum on the tempering of quenched 0.1C steels (Irving and Pickering, *JISI*, 1960, **194**, 137)

8.4.3 Nucleation and growth of alloy carbides

The dispersions of alloy carbides which occur during tempering can be very complex, but some general principles can be discerned which apply to a wide variety of steels. The alloy carbides can form in at least three ways:
(1) *In-situ* nucleation at pre-existing cementite particles – it has been shown that the nuclei form on the interfaces between cementite particles and the ferrite. As they grow, carbon is provided by the adjacent cementite which gradually disappears.
(2) By separate nucleation within the ferrite matrix – usually on dislocations inherited from the martensitic structure.
(3) At grain boundaries and subboundaries – these include the former

austenite boundaries, the original martensitic lath boundaries (now ferrite), and the new ferrite boundaries formed by coalescence of subboundaries, or by recrystallization.

In-situ nucleation at pre-existing cementite particles is a common occurrence but, because these particles are fairly widely spaced at temperatures above 500° C, the contribution of this type of alloy carbide nucleation to strength is very limited. Fig. 8.11a shows, in a 4% molybdenum steel tempered $4\frac{1}{2}$ hours at 550° C, the relatively coarse Widmanstätten precipitation of Fe_3C, which at this stage has largely

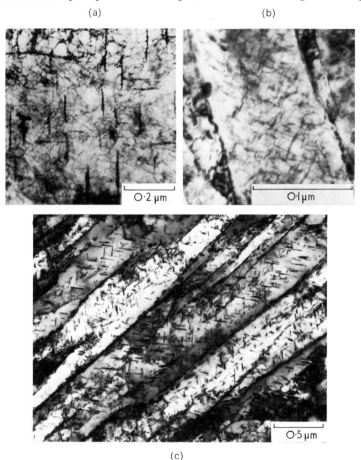

Fig. 8.11 Tempering of an Fe-4Mo-0.2C steel. Thin-foil EMs: a, $4\frac{1}{2}$ h at 550°C. Cementite transforming *in situ* to Mo_2C and start of nucleation on dislocations (Raynor *et al.*); b, 5h at 600°C, Mo_2C precipitation on dislocations within a former martensite lath (Irani); c, 30 min at 700°C Mo_2C precipitation in ferrite laths inherited from martensite. M_6C precipitation at lath boundaries (Irani)

transformed to fine Mo_2C particles. These are readily identified by dark field microscopy. On further tempering, the positions of the original cementite particles are indicated by small necklaces of alloy carbides which tend to be coarser than the matrix precipitation.

Fig. 8.11a also illustrates the dislocation network characteristic of tempered steels and inherited from the martensite, although there has been considerable rearrangement and reduction in dislocation density. Dark field electron microscopy reveals that these dislocations are the sites for very fine precipitation of the appropriate alloy carbide. On further ageing the particles are more readily resolved, e.g. in Mo steels as a Widmanstätten array, comprising Mo_2C rodlets lying along $\langle 001 \rangle_\alpha$ directions. Fig. 8.11b illustrates this stage in a single martensitic lath. Heavier precipitation is evident at the lath boundaries.

The nucleation of carbides at the various types of boundary is to be expected because these are energetically favourable sites which provide paths for relatively rapid diffusion of solute. Consequently the ageing process is usually more advanced in these regions and the precipitate is more massive (Fig. 8.11c). In many alloy steels, the first alloy carbide to form is not the final equilibrium carbide and, in some steels, as many as three alloy carbides can form successively. In these circumstances, the equilibrium alloy carbide frequently nucleates first in the grain boundaries, grows rapidly and eventually completely replaces the Widmanstätten non-equilibrium carbide within the grains. This is illustrated in Fig. 8.11c for a 4 Mo steel tempered 30 min at 700°C, in which M_6C equi-axed particles are growing at the grain boundaries but Widmanstätten Mo_2C is still visible within the grains. It is interesting to note that the structure still possesses the lath-shaped ferrite grains inherited from the martensite. Recrystallization occurs after longer times at 700°C.

8.4.4 Tempering of steels containing vanadium

Vanadium is a strong carbide former and, in steel with as little as 0.1% V, the face-centred cubic vanadium carbide VC is formed. It is often not of stoichiometric composition, being frequently nearer V_4C_3, but with other elements in solid solution within the carbide. Normally, this is the only vanadium carbide formed in steels, so the structural changes during tempering of vanadium steels are relatively simple.

Vanadium carbide forms as small platelets, initially less than 5 nm across and not more than 1 nm thick. These form within the ferrite grains on dislocations (Fig. 8.12a) in the range 550–650°C, and produce a marked secondary hardening peak. There is a well-defined orientation relationship (Baker/Nutting) with the ferrite matrix: $\{100\}_{VC}//\{100\}_\alpha$, $\langle 100 \rangle_{VC}//\langle 110 \rangle_\alpha$. In the early stages of precipitation at 550°C, the particles are coherent with the matrix, there being only a 3% misfit between $[010]_\alpha$ and $[011]_{VC}$. However, at 700°C, the platelets coarsen rapidly

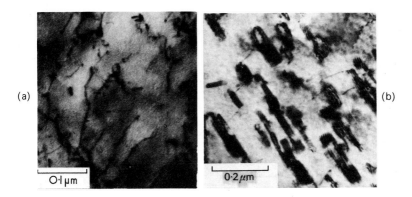

Fig. 8.12 Fe-1V-0.2C quenched and tempered. Thin-foil EMs: a, 72h at 550°C, VC nucleation on dislocations (Raynor); b, 50h at 700°C. Plates of VC (Irani)

(Fig. 8.12b) and begin to spheroidize. However, the original martensite laths can still be recognised, and are only replaced by equi-axed ferrite grains after long periods at 700°C.

Many steels containing vanadium, e.g. $\frac{1}{2}$Cr$\frac{1}{2}$Mo$\frac{1}{4}$V, 1Cr$\frac{1}{4}$V, 3Cr1Mo$\frac{1}{4}$V, 1Cr1Mo$\frac{3}{4}$V, will exhibit extensive vanadium carbide precipitation on tempering, because of the stability of this carbide, not only with respect to cementite but also the several chromium carbides and molybdenum carbide (see Fig. 4.5). Because of its ability to maintain a fine carbide dispersion, even at temperatures approaching 700°C, vanadium is an important constituent of steels for elevated temperature service.

8.4.5 Tempering of steels containing chromium

In chromium steels, two chromium carbides are very often encountered: Cr_7C_6 (trigonal) and $Cr_{23}C_6$ (complex cubic). The normal carbide sequence during tempering is

$$\text{Matrix} \rightarrow (\text{FeCr})_3\text{C} \rightarrow Cr_7C_3 \rightarrow Cr_{23}C_6$$

While this sequence occurs in higher chromium steels, below about 7% Cr, $Cr_{23}C_6$ is absent unless other metals such as molybdenum are present. Chromium is a weaker carbide former than vanadium, which is illustrated by the fact that Cr_7C_3 does not normally occur until the chromium content of the steel exceeds 1% at a carbon level of about 0.2%.

In steels up to 4% Cr, the transformation from Fe_3C to Cr_7C_6 occurs mainly by nucleation at the Fe_3C/ferrite interfaces. Steels up to 9% Cr do not show secondary hardening peaks in tempering curves (Fig. 8.13). However these curves do exhibit plateaus at the higher chromium contents, which are associated with the precipitation of Cr_7C_3. Chromium diffuses

158 *Steels—Microstructure and Properties*

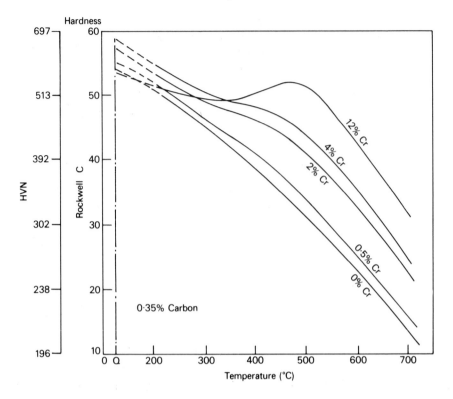

Fig. 8.13 The effect of chromium on the tempering of a 0.35C steel (Bain and Paxton, *The Alloying Elements in Steel*, 2nd edition, ASM, 1961)

more rapidly in ferrite than most metallic alloying elements, with the result that Cr_7C_3 is detected during tempering at temperatures as low as 500°C, and in comparison with vanadium carbide, chromium carbide coarsens rapidly. Thus, in a 2% Cr 0.2% C steel, continuous softening will normally occur on tempering between 500 and 700°C, although addition of other alloying elements, e.g. Mo, can reduce the rate of coarsening of Cr_7C_3.

In contrast, a 12% Cr steel will exhibit secondary hardening in the same temperature range (Fig. 8.13) due to precipitation of Cr_7C_3. Additionally, $Cr_{23}C_6$ nucleates at about the same time but at different sites, particularly former austenite grain boundaries and at ferrite lath boundaries. This precipitate grows at the expense of the Cr_7C_3 which eventually disappears from the microstructure, at which stage the steel has completely over-aged. This transition from Cr_7C_3 to $Cr_{23}C_6$ in high chromium steels is by separate nucleation and growth. Further alloying additions can promote one or other of these carbide reactions, e.g. addition of tungsten encourages formation of $Cr_{23}C_6$ by allowing it to nucleate faster, while vanadium tends to stabilize Cr_7C_3. In doing so, it decreases the rate of release into solution of chromium and carbon needed for the growth of $Cr_{23}C_6$. Clearly, vanadium would be a preferred addition to tungsten, if a fine stable chromium carbide dispersion is needed in the temperature range 550–650°C.

8.4.6 Tempering of steels containing molybdenum and tungsten

When molybdenum or tungsten is the predominant alloying element in a steel, a number of different carbide phases are possible, but for composition between 4 and 6 wt% of the element the carbide sequence is likely to be:

$$Fe_3C \rightarrow M_2C \rightarrow M_6C$$

The carbides responsible for the secondary hardening in both the case of tungsten and molybdenum are the isomorphous hexagonal carbides Mo_2C and W_2C, both of which, in contrast to vanadium carbide, have a well-defined rodlet morphology (Fig. 8.14a). When formed in the matrix, M_2C adopts a Widmanstätten distribution lying along $\langle 100 \rangle_\alpha$ directions. In

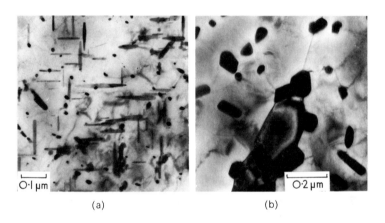

Fig. 8.14 Fe-6W-0.23C quenched and tempered (Davenport). Thin-foil EMs: a, 100h at 600°C. W_2C needles along $\langle 001 \rangle$ some irregular particles of M_6C; b, 26h at 700°C. Massive M_6C

molybdenum steels, peak hardness occurs after about 25 hours at 550°C, when the rods are about 10–20nm long and 1–2nm in diameter. The orientation relationship is:

$(0001)_{M_2C}//(011)_\alpha$

$[11\bar{2}0]_{M_2C}//[100]_\alpha$ (rod growth direction)

M_2C also nucleates at former austenite and ferrite lath boundaries. As in the case of vanadium steels, the M_2C precipitate nucleates both on dislocations in the ferrite, and at the Fe_3C/ferrite interfaces, but the secondary hardening arises primarily from the dislocation-nucleated dispersion of M_2C.

On prolonged tempering at 700°C, the complex cubic M_6C forms predominantly at grain boundaries as massive particles which grow quickly, while the M_2C phase goes back into solution. The equilibrium microstructure is equi-axed ferrite with coarse M_6C in the form of faceted particles at grain boundaries, and plates, illustrated in Fig. 8.14b for a 6% tungsten steel tempered 26 hours at 700°C.

For similar atomic concentrations, the secondary hardening response in the case of tungsten steels is less than that of molybdenum steels. The M_2C dispersion in the former case is coarser, probably because the slower diffusivity of tungsten allows a coarsening of the dislocation network prior to being pinned by the nucleation of M_2C particles.

At lower concentrations of tungsten and molybdenum ($\frac{1}{2}$–2%), two other alloy carbides are interposed in the precipitation sequences, i.e. the complex cubic $M_{23}C_6$ and the orthorhombic M_aC_b, probably Fe_2MoC. These carbides are found as intermediate precipitates between M_2C and M_6C.

8.4.7 Complex alloy steels

The presence of more than one carbide forming element can complicate the precipitation processes during tempering. In general terms, the carbide phase which is the most stable thermodynamically will predominate, but this assumes that equilibrium is reached during tempering. This is clearly not so at temperatures below 500–600°C. The use of pseudo-binary diagrams for groups of steels, e.g. Cr–V, Cr–Mo, can be a useful guide to carbide phases likely to form during tempering (see Chapter 4, Section 4.2). The sequence of precipitation for a particular composition can be approximated to by drawing a line from the origin of the diagram, e.g. Fig. 4.6, to the composition of interest. The phase fields passed through would normally be those encountered in tempering, but the exact conditions cannot be forecast from such data.

Certain strong carbide formers, notably niobium, titanium and vanadium, have effects on tempering out of proportion to their concentration. In concentrations of 0.1 wt% or less, provided the tempering temperature is high enough, i.e. 550–650°C, they combine preferentially with part of the carbon and, in addition to the major carbide phase, e.g. Cr_7C_3, Mo_2C, they form a separate, very much finer dispersion, more resistant to over-ageing (Fig. 8.15). This secondary dispersion can greatly augment the secondary hardening reaction, illustrating the importance of these strong carbide forming elements in achieving high strength levels, not only at room temperature but also at elevated temperatures, where creep resistance is often an essential requirement.

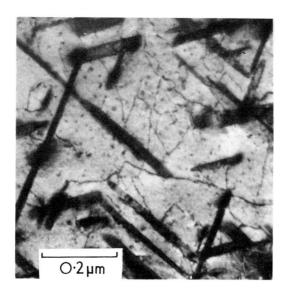

Fig. 8.15 Fe-4Mo-0.1Nb-0.2C steel tempered 6h at 700°C. Coarse needles of Mo_2C in ferrite and fine particles of NbC on dislocations (Irani). Thin-foil EM

8.4.8 Mechanical properties of tempered alloy steels

A wide range of mechanical properties is obtainable by tempering alloy steels between 200°C and 700°C. A typical example is shown in Fig. 8.16 for a steel containing $1\frac{1}{2}$Ni 1%Cr 0.25%Mo 0.4%C (En 24), the tensile strength of which can be varied from 1800 down to 900 MNm^{-2} by tempering at progressively higher temperatures. The ductility of the steel improves as the tensile strength falls. However, there is a ductility minimum around 275–300°C, which is often observed in plain carbon and lower alloy steels. This has been attributed to the conversion of retained austenite to bainite, but it is more likely to be the result of the formation of thin cementite films, as a result of the transformation of austenite at the interlath boundaries. At higher temperatures, these films spheroidize and the toughness improves.

To obtain really high strength levels in tempered steels ($\sim 1500\,MNm^{-2}$), it is usual to temper at low temperatures, i.e. 200–300°C, when the martensite is still heavily dislocated and the main strengthening dispersion is cementite or ε-iron carbide. Alloy steels, when tempered in this range, not only provide very high tensile strengths with some ductility but are also superior to plain carbon steels, as shown in Fig. 8.17. It is clear from Fig. 8.17a that the carbon content has a large influence on the strength. The alloying elements refine the iron carbide dispersion and, as the carbon content is raised, the dispersion becomes more dense and, therefore, more

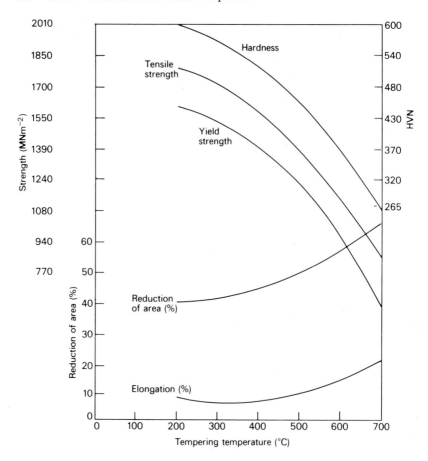

Fig. 8.16 Mechanical properties of En 24 (1.5Ni-1Cr-0.25Mo-0.4C steel) as a result of tempering for 1h (Thelning, *Steel and its Heat Treatment*, Bofors Handbook, Butterworths, 1975)

effective. The toughness decreases with increasing strength, as shown in Fig. 8.17b. However, alloying elements very substantially improve the toughness, when compared with plain carbon steels of similar strength levels. When molybdenum is present in the steel, the toughness is increased further as the scatter bands indicate. This effect of alloying elements is again attributed to the breakdown of carbide films at grain and martensite lath boundaries. These films are particularly less noticeable in steels containing molybdenum.

Alloy steels which exhibit secondary hardening can provide high strength levels on tempering between 500 and 700°C, with better ductility than that obtained at lower tempering temperatures. However, one of their main advantages is that, once a high strength level is reached by means of an alloy

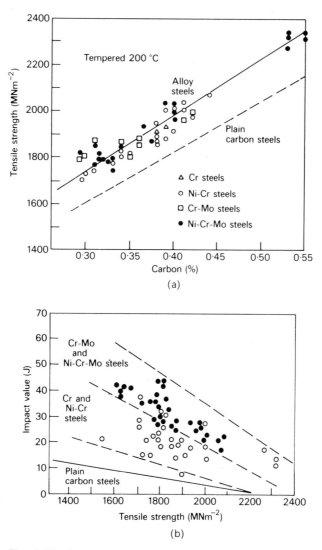

Fig. 8.17 Comparison of mechanical properties of plain carbon and alloy steels tempered at 200°C (Irving and Pickering, *JISI*, 1960, **194**, 137): a, effect of C on tensile strength; b, relation between tensile strength and impact value. Note beneficial effect of Mo.

carbide dispersion formed between 550 and 650°C, this structure will be relatively stable at temperatures up to 500°C. Therefore, the steels are suitable for use under stress at elevated temperatures. A typical example is given in Fig. 8.18 of a 12%Cr 1%Ni 0.2%C stainless steel, which can be quenched to martensite and then tempered to give a fine dispersion of

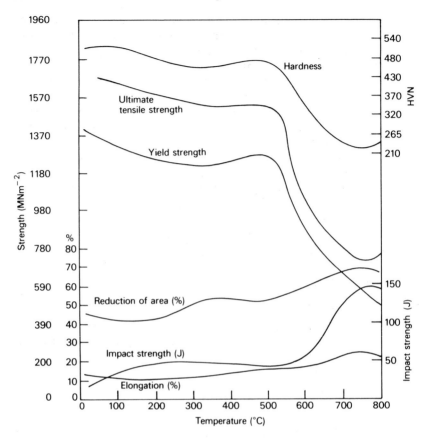

Fig. 8.18 Effect of tempering for 1h on the mechanical properties of a 12Cr-1Ni-0.2C stainless steel. Typical results for 50 mm diameter bars, oil-quenched (Thelning, *Steel and its Heat Treatment*, Bofors Handbook, Butterworths, 1975)

chromium carbides in a ferritic matrix. The strength is well-maintained up to the secondary hardening peak at 500°C, and is combined with a reasonable level of ductility. This type of steel is tempered to between 700 and 1000 MNm^{-2} yield stress and is frequently used in steam and gas turbines, but can also be used for constructional purposes where lower temperatures are involved. Further improvements in mechanical properties at elevated temperatures can be obtained by addition of small concentrations of stronger carbide formers, e.g. molybdenum (2%) and vanadium (0.25%).

8.5 Maraging steels

It has been shown that precipitation of alloy carbides in tempered martensite gives rise to age-hardening, usually referred to as secondary hardening. There is no reason why other finely divided phases cannot be used for a similar purpose and, in fact, an important group of high alloy steels, the maraging steels, reach high strength levels by the precipitation of various intermetallic compounds.

Carbide precipitation is practically eliminated by the use of low carbon compositions, and the steels contain between 18 and 25% nickel so that, on quenching from the austenitic condition, they form a soft but heavily dislocated martensite, often referred to as massive martensite. The high nickel content lowers the M_s to around 150°C, but on reheating the martensite there is considerable hysteresis, so that austenite is not reformed until the steel is held between 500 and 600°C. At somewhat lower temperatures, i.e. 400–500°C, precipitation of intermetallic phases takes place, accelerated by the influence of the high dislocation density on the diffusion of substitutional solute atoms. Elements such as molybdenum and titanium are necessary additions, which result in the precipitation of Ni_3Mo, Ni_3Ti and the Laves phase, Fe_2Mo. Cobalt is also a useful alloying element as it reduces the solubility of molybdenum in the matrix and this increases the volume fraction of molybdenum-rich precipitate.

The precipitate reactions can lead to very high volume fractions of precipitate, and thus to the achievement of high strength levels (Equations (2.10) and (2.11)). For example, a steel with 18–19 Ni, 8.5–9.5 Co, 4.5–5 Mo and 0.5–0.8 Ti can be heat treated to give a yield stress around 2000 MNm^{-2}. However, the important point is that these high strength levels are accompanied by good ductility and toughness.

Further reading

Roberts, C. S., Averbach, B. L. and Cohen, M., *Trans. ASM.*, **45**, 576, 1953

Kurdjumov, G., *J.I.S.I.*, **195**, 26, 1960

Winchell, P. G. and Cohen, M., *Trans. ASM.*, **55**, 347, 1962

Irvine, K. J. and Pickering, F. B., High Strength 12% Chromium Steels, in Iron and Steel Institute Special Report No. 86, 34, 1964

Woodhead, J. H. and Quarrell, A. G., *The Role of Carbides in Low Alloy Creep-Resisting Steels*, Climax Molybdenum Co., 1965

Speich, G. R. and Clark, J. B. (eds), *Precipitation from Iron-Base Alloys*, Gordon and Breach, 1965

Speich, G. R., Tempering of plain carbon martensite. *Trans. AIME*, **245**, 2553, 1969

Nutting, J., Physical metallurgy of alloy steels. *J.I.S.I.*, **207**, 872, 1969

Honeycombe, R. W. K., *Structure and Strength of Alloy Steels*, Climax Molybdenum Co., 1973

Thomas, G., Retained austenite and tempered martensite embrittlement. *Met. Trans.*, **9**, 439, 1978

Pickering, F. B., *Physical Metallurgy and the Design of Steels*, Applied Science Publishers, 1978

9
Thermomechanical treatment of steels

9.1 Introduction

Thermomechanical treatment involves the simultaneous application of heat and a deformation process to an alloy, in order to change its shape and refine the microstructure. Thus, hot-rolling of metals, a well-established industrial process, is a thermomechanical treatment which plays an important part in the processing of many steels from low carbon mild steels to highly alloyed stainless steels. The traditional fabrication route involves the casting of ingots varying in size from 1 tonne to 50 tonne, which are soaked at very high temperatures (1200–1300°C), then progressively hot rolled to billets, bars and sheet. This leads to the breaking down of the original coarse cast structure by repeated recrystallization of the steel while in the austenitic condition, and by the gradual reduction of inhomogeneities of composition caused by segregation during casting. Also, the inevitable non-metallic inclusions, i.e. oxides, silicates, sulphides, are broken up, some deformed, and distributed throughout the steel in a more uniform manner.

The hot-rolling process has gradually become a much more closely-controlled operation, and is being increasingly applied to low alloy steels with compositions carefully chosen to provide optimum mechanical properties when the hot deformation is complete. This process, in which the various stages of rolling are temperature-controlled, the amount of reduction in each pass is predetermined and the finishing temperature is precisely defined, is called *controlled rolling* and is now of greatest importance in obtaining reliable mechanical properties in steels for pipelines, bridges, and many other engineering applications.

At the other end of the scale, in more highly alloy steels, it is possible to subject the steels to heavy deformations in the metastable austenitic conditions prior to transformation to martensite. This process, *ausforming*, allows the attainment of very high strength levels combined with good toughness and ductility.

9.2 Controlled rolling of low alloy steels

9.2.1 General

Before World War II, strength in hot-rolled low alloy steels was achieved by the addition of carbon up to 0.4% and manganese up to 1.5%, giving yield stresses of 350–400 MNm^{-2}. However, such steels are essentially ferrite pearlite aggregates, which do not possess adequate toughness for many modern applications. Indeed, the toughness, as measured by the ductile/brittle transition decreases dramatically with carbon content, i.e. with increasing volume of pearlite in the steel (Fig. 9.1). Furthermore, with the introduction of welding as the main fabrication technique, the high carbon contents led to serious cracking problems, which could only be eliminated by the use of lower carbon steels. The great advantage of producing in these steels a fine ferrite grain size soon became apparent (see Section 2.5), so controlled rolling in the austenitic condition was gradually introduced to achieve this. Fine ferrite grain sizes in the finished steel were found to be greatly expedited by the addition of small concentrations (< 0.1 wt%) of grain refining elements such as niobium, titanium and vanadium, and also aluminium. On adding such elements to steels with 0.03–0.08%C and up to 1.5% Mn, it became possible to produce fine-grained material with yield strengths between 450 and 550 MNm^{-2}, and with ductile/brittle transition temperatures as low as $-70°$C. Such steels are now referred to as high strength low alloy steels (HSLA), or micro-alloyed steels. This progress, from the relatively low strength of ordinary mild steel (220–250 MNm^{-2}) in a period of twenty years represents a major metallurgical development, the importance of which, in engineering applications, cannot be overstated.

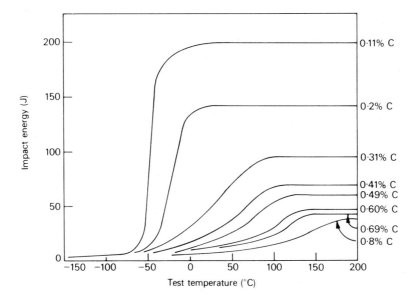

Fig. 9.1 Effect of carbon content on the impact transition temperature curves of ferrite/pearlite steels (Pickering, In: *Micro-alloying 75*, Union Carbide Corporation, 1975)

9.2.2 Grain size control during controlled rolling

The primary grain refinement mechanism in controlled rolling is the recrystallization of austenite during hot deformation, known as *dynamic recrystallization*. This process is clearly influenced by the temperature and the degree of deformation which takes place during each pass through the rolls. However, in austenite devoid of second phase particles, the high temperatures involved in hot rolling lead to marked grain growth, with the result that grain refinement during subsequent working is limited.

The situation is greatly improved, if fine particles are introduced into the austenitic matrix. The particles are usually found on grain boundaries, because an interaction takes place between the particles and the boundary. A short length of grain boundary is replaced by the particle and the interfacial energy ensures a stable configuration. When the grain boundary attempts to migrate away from the particles, the local energy increases and thus a drag is exerted on the boundary by the particles.

The theory of boundary pinning by particles has already been referred to in Chapter 8. Equation (8.1) defines the critical size of particle below which pinning is effective. Clearly, the control of grain size at high austenitizing temperatures requires as fine a grain boundary precipitate as possible, and one which will not dissolve completely in the austenite, even at the highest working temperatures (1200–1300°C). The best grain refining elements are very strong carbide and nitride formers, such as niobium, titanium and vanadium, also aluminium which forms only a nitride. As both carbon and nitrogen are present in control-rolled steels, and as the nitrides are even more stable than the carbides (see Fig. 4.5, Chapter 4), it is likely that the most effective grain refining compounds are the respective carbo-nitrides, except in the case of aluminium nitride.

Equally important is the degree of solubility that such stable compounds have in austenite. It is essential that there is sufficient solid solubility at the highest austenitizing (soaking) temperatures to allow fine precipitation to occur during controlled rolling at temperatures which decrease as rolling proceeds. The solubility products (in atomic percent) of several relevant carbides and nitrides have been shown in Fig. 4.12 as a function of the reciprocal of the temperature. All of these compounds have a small but increasing solubility in the critical temperature range (900–1300°C) (Fig. 9.2). In contrast, the carbides of chromium and molybdenum have much higher solubilities, which ensure that they will normally go completely into solution in the austenite, if the temperature is high enough, and will not precipitate until the temperature is well below the critical range for grain growth. Data from another source[†] has provided the following equations for solubilities expressed in weight percent as a function of absolute temperature:

[†] Irvine, K. J. *et al.*, *JISI*, 1967, **205**, 161

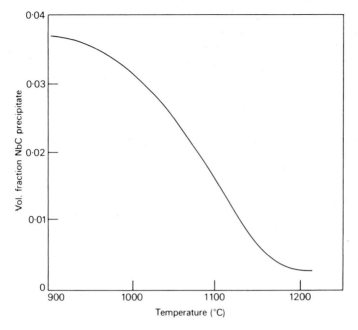

Fig. 9.2 Solubility curve for NbC in a steel with 0.15C 1.14Mn-0.04Nb (Hoogendoorn and Spanraft, In: *Micro-alloying 75*, Union Carbide Corporation, 1975)

$\log_{10}[\text{Al}][\text{N}] = -6770/T + 1.03$
$\log_{10}[\text{V}][\text{N}] = -8330/T + 3.46$
$\log_{10}[\text{Nb}][\text{C}] = -6770/T + 2.26$
$\log_{10}[\text{Ti}][\text{C}] = -7000/T + 2.75$

The compositional changes possible are many, so discussion will be limited to general principles which apply equally, whichever compound is the effective grain refiner in a given steel. While grain growth at the highest austenitizing temperatures may be restricted to some extent by a residual dispersion, the main refinement is achieved during rolling as the temperature progressively falls, and fine carbo-nitrides are precipitated from the austenite. These new precipitates will:
(1) Increase the strain, for a given temperature, at which recrystallization will commence (Fig. 9.3)
(2) Restrict the movement of recrystallized grain boundaries.

It should be borne in mind that the austenite may recrystallize several times during a controlled rolling schedule and the total effect of this will be a marked austenite grain refinement by the time the steel reaches the γ/α transformation temperature (Fig. 9.3). In the later stages of austenite deformation, at the lower temperatures, recrystallization may not occur,

170 Steels—Microstructure and Properties

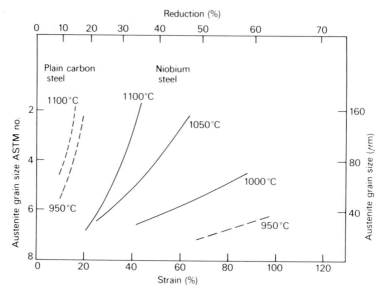

Fig. 9.3 Critical strain needed to complete recrystallization of austenite as a function of deformation temperature and grain size. Comparison of Nb steel with plain carbon steel (Tanaka et al., In: Micro-alloying 75, Union Carbide Corporation, 1975)

with the result that deformed austenite grains elongated and flattened by rolling may transform directly to ferrite. In the final stages of controlled rolling, austenite grain growth can be further suppressed by rapid cooling from the finishing temperature, which allows the γ/α transformation to take place subcritically, i.e. below Ar_1, in austenite which is still deformed. It is becoming common practice now to continue rolling through the γ/α transformation, and even into the fully ferritic region. Such treatments lead to finer grain sizes, and higher yield stresses in the finished product (Fig. 9.4), but impose much higher load factors on the rolling mills.

As a result of the combined use of controlled rolling and fine dispersions of carbo-nitrides in low alloy steels, it has been possible to obtain ferrite grain sizes between 5 and 10 μm in commercial practice. Laboratory tests have achieved grain sizes approaching 1 μm, which would appear to be a practical limit using this approach. The Hall-Petch relationship between grain size and yield strength, which was discussed in Chapter 2, is very relevant to micro-alloyed steels and, in fact, linear plots are obtained for the yield stress against $d^{-\frac{1}{2}}$ (Fig. 9.5). Addition of 0.05–0.09% Nb to a plain carbon steel refines the ferrite grain size, allowing it to be reduced to below 5 μm ($d^{-\frac{1}{2}} = 14$), with a consequent substantial increase in yield strength. The displacement of ~ 100 MNm^{-2} between the C-Mn and C-Mn-Nb curves arises from dispersion strengthening due to NbC. This is further illustrated in the two lower curves of Fig. 9.6, which were obtained

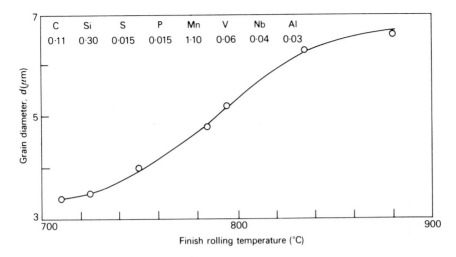

Fig. 9.4 Effect of finish rolling temperature on final ferrite grain size of a micro-alloyed steel (after McKenzie, *Proc. Rosenhain Centenary Conference*, Royal Society, 1976)

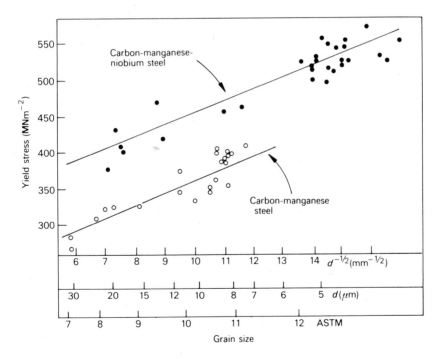

Fig. 9.5 Effect of grain size on yield stress of a carbon-manganese-niobium steel (Le Bon and Saint Martin, In: *Micro-alloying 75*, Union Carbide Corporation, 1975)

172 *Steels—Microstructure and Properties*

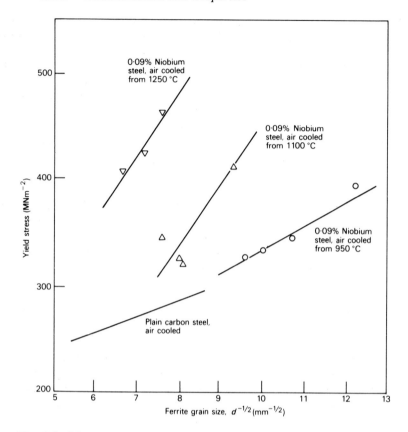

Fig. 9.6 Effect of austenitizing temperature on the yield strength of a 0.1C-0.6Mn-0.09Nb steel (Gladman *et al.*, In: *Micro-alloying 75*, Union Carbide Corporation, 1975)

from specimens austenitized at 950°C prior to air cooling. If, however, progressively higher austenitizing temperatures are used, e.g. 1100°C and 1250°C followed by air cooling, the resulting curves, although still linear, have much steeper slopes, indicating a marked increase in yield strength for a particular grain size. This large increment in strength is due to the precipitation of NbC during cooling, following its solution at the higher austenitizing temperatures.

9.2.3 Dispersion strengthening during controlled rolling

The solubility data implies that, in a micro-alloyed steel, carbides and carbo-nitrides of Nb, Ti and V will precipitate progressively during controlled rolling as the temperature falls. While the primary effect of these fine dispersions is to control grain size, dispersion strengthening will take place. The strengthening arising from this cause will depend both on the particle size r, and the interparticle spacing Λ which is determined by the volume fraction of precipitate (Equation (2.10)). These parameters will depend primarily on the type of compound which is precipitating, and that is determined by the micro-alloying content of the steel. However, the maximum solution temperature reached and the detailed schedule of the controlled rolling operation are also important variables.

It is now known, not only that precipitation takes place in the austenite, but that further precipitation occurs during the transformation to ferrite. The precipitation of niobium, titanium and vanadium carbides has been shown to take place progressively as the interphase boundaries move through the steel. This is the interphase precipitation discussed in Chapter 4, Section 4.4.3. As this precipitation is normally on an extremely fine scale occurring between 850 and 650°C, it is likely to be the major contribution to the dispersion strengthening. In view of the higher solubility of vanadium carbide in austenite, the effect will be most pronounced in the presence of this element, with titanium and niobium in decreasing order of effectiveness. If the rate of cooling through the transformation is high, leading to the formation of supersaturated acicular ferrite, the carbides will tend to precipitate within the grains, usually on the dislocations which are numerous in this type of ferrite.

In arriving at optimum compositions of micro-alloyed steels, it should be borne in mind that the maximum volume fraction of precipitate which can be put into solid solution in austenite at high temperatures is achieved by use of stoichiometric compositions. For example, if titanium (atomic weight 47.9) is used, it will combine with approximately one quarter its weight of carbon (atomic weight 12), so that for a 0.025 wt % C steel, 0.10 wt % of Ti will provide carbide of the stoichiometric composition. In Fig. 9.7 the stoichiometric line for TiC is shown superimposed on the solubility curves for titanium carbide at 1100° C, 1200° C and 1300° C. If the precipitation in steels with 0.10 % titanium cooled from 1200°C is considered, at low carbon contents, i.e. to the left of the stoichiometric line, the carbide fraction is limited by the carbon content, i.e. Zone A, lower diagram. For carbon contents between the stoichiometric line and the solubility line at 1200°C, the full potential volume fraction of fine TiC will form on cooling (Zone B). When the carbon content exceeds the solubility limit (> 0.10 wt %), the titanium is progressively precipitated at 1200°C as coarse carbide, thus reducing the amount of titanium available to combine with carbon to form fine TiC during cooling. As coarse carbide particles are ineffective in controlling grain growth, it is highly desirable to have steel compositions which avoid their formation. It also follows from Fig. 9.7 that high austenitizing temperatures are essential to obtain full benefit from the precipitation of finely divided carbide phases.

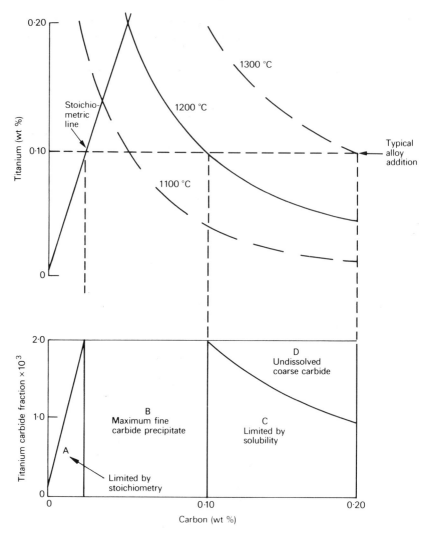

Fig. 9.7 Effect of stoichiometry on the precipitation of TiC in a micro-alloyed steel (Gladman et al., In: *Micro-alloying 75*, Union Carbide Corporation, 1975)

9.2.4 Strength of micro-alloyed steels—an overall view

In modern control-rolled micro-alloyed steels, there are at least three strengthening mechanisms which contribute to the final strength achieved. The relative contribution from each is determined by the composition of the steel and, equally important, the details of the thermomechanical treatment to which the steel is subjected. The several strengthening contribuions for steels with 0.2% carbon, 0.2% silicon, 0.15% vanadium and 0.015% nitrogen as a function of increasing manganese content are shown schematically in Fig. 9.8. Firstly, there are the solid solution strengthening increments from manganese, silicon and uncombined nitrogen. Secondly,

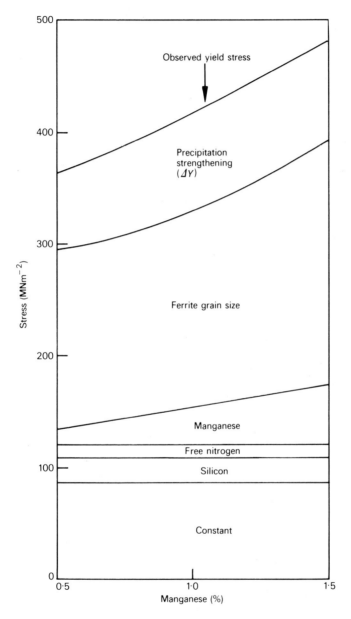

Fig. 9.8 The contributions to strength in a 0.2C-0.15V steel as a function of Mn content (Gladman et al., In: *Micro-alloying 75*, Union Carbide Corporation, 1975)

the grain size contribution to the yield stress is shown as a very substantial component, the magnitude of which is very sensitive to the detailed thermomechanical history. Finally, a typical increment for dispersion strengthening is shown. The total result is a range of yield strengths between about 350 and 500 MNm^{-2}. In this particular example, the steel was normalized (air cooled) from 900°C, but had it been control rolled down to 800°C or even lower, the strength levels would have been substantially raised.

The effect of the finishing temperature for rolling is important in determining the grain size and, therefore, strength level reached for a particular steel. It is now becoming common to roll through the transformation into the completely ferritic condition, and so obtain fine subgrain structures in the ferrite, which provide an additional contribution to strength. Alternatively, the rolling is finished above the γ/α transformation, and the nature of the transformation is altered by increasing the cooling rate. Slow rates of cooling obtained by coiling at a particular temperature will give lower strengths than rapid rates imposed by water spray cooling following rolling. The latter route can change the ferrite from equi-axed to Widmanstätten with a much higher dislocation density. The result is a steel with improved mechanical properties and, in many cases, the sharp yield point can be suppressed. This has practical advantages in fabrication of sheet steel, e.g. pipe manufacture, where a continuous stress-strain curve is preferred.

9.3 Dual phase steels

The high strength low alloy steels described in Section 9.2 give improved strength to weight ratios over ordinary standard steels. However, they are not readily formed, e.g. by cold pressing and related techniques. This has led to difficulties in, for example, the car industry in the USA where legislation on safety and fuel economy is hastening a trend towards the use of higher strength steels for many parts. Work initiated in response to this need has shown that low alloy steels, typically containing manganese and silicon, can exhibit both high strength and very good formability if they are first heat treated to produce a matrix of ferrite with islands of martensite (10–20 % by volume). These steels are referred to as dual phase low alloy (DPLA) steels. They exhibit continuous yielding, i.e. no sharp yield point, and a relatively low yield stress (300–350 MNm^{-2}) together with a rapid rate of work hardening and high elongations ($\sim$ 30 %) which gives excellent formability. As a result of the work hardening, the yield stress in the final formed product is as high as in HSLA steels (500–700 MNm^{-2}). The simplest steels in this category contain 0.08–0.2 % C, 0.5–1.5 % Mn, but steels microalloyed with vanadium are also suitable, while small additions of Cr (0.5 %) and Mo (0.2–0.4 %) are frequently used.

The simplest way of achieving a duplex structure is to use intercritical annealing in which the steel is heated to the $(\alpha + \gamma)$ region between Ac_1 and Ac_3 and held, typically, at 790° C for several minutes to allow small regions of austenite to form in the ferrite. As it is essential to transform these regions to martensite, recooling must be rapid or the austenite must have a high hardenability. This can be achieved by adding between 0.2 and 0.4 Mo to a steel already containing 1.5% manganese. The required structure can then be obtained by air cooling after annealing.

To eliminate an extra heat treatment step, dual phase steels have now been developed which can be given the required structure during cooling after controlled rolling. Typically, these steels have additions of 0.5Cr and 0.4 Mo. After completion of hot rolling around 870° C, the steel forms approximately 80% ferrite on the water cooled run-out table from the mill. The material is then coiled in the metastable region (510–620° C) below the pearlite/ferrite transformation and, on subsequent cooling, the austenite regions transform to martensite.

9.4 Ausforming

9.4.1 General

The process known as *ausforming* or low temperature thermomechanical treatment (LTMT), which was first described by Harvey and later by Lips and Van Zuilen, involves the deformation of austenite in the metastable bay between the ferrite and bainite curves of the *TTT* diagram. The treatment is shown schematically in Fig. 9.9a. A steel with a sufficiently developed metastable austenite bay is quenched from the austenitizing temperature to this region, where substantial deformation is carried out, without allowing transformation to take place. The deformed steel is then transformed to martensite during quenching to room temperature, and the appropriate balance of mechanical properties achieved by subsequent tempering. This ausforming treatment can be contrasted with a high temperature thermomechanical treatment (HTMT), where the deformation is carried out in the stable austenite region (Fig. 9.9b), usually above Ae_3 prior to quenching to form martensite. In a third process, *isoforming* (Fig. 9.9c), the steel is deformed in the metastable austenite region, but the deformation is continued until the transformation is complete at the intermediate temperature. The steel can then be slowly cooled to room temperature.

The ausforming process needs careful control to be successful and usually involves very substantial deformation. However, the attraction is that with appropriate steels dramatic increases in strength are achieved without adverse effect on ductility and toughness. Typically, a 4.7% Cr, 1.5% Mo, 0.4% V, 0.34C steel has a tensile strength of about 2000 MNm^{-2} after conventional quenching and tempering, whereas after ausforming the strength can be over 3000 MNm^{-2}.

178 *Steels—Microstructure and Properties*

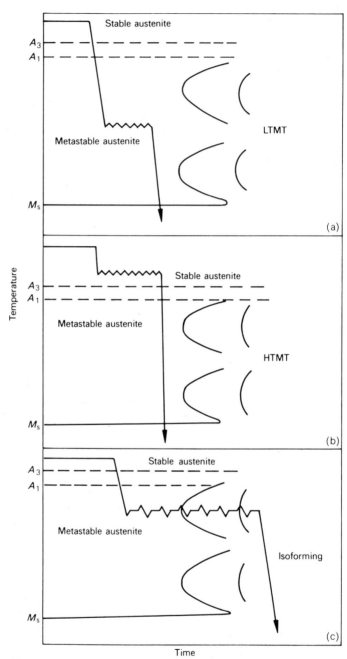

Fig. 9.9 Schematic diagrams of some thermochemical treatments: a, ausforming–low temperature mechanical treatment; b, high temperature mechanical treatment; c, isoforming

9.4.2 The composition of the steel

Steels, in which austenite transforms rapidly at subcritical temperatures, are not suitable for ausforming. It is necessary to add alloying elements which develop a deep metastable austenite bay by displacing the *TTT* curve to longer transformation times. The most useful elements in this respect are chromium, molybdenum, nickel and manganese, and allowance must be made for the fact that deformation of the austenite accelerates the transformation. Consequently, it is necessary to have sufficient alloying element present to slow down the reaction and avoid the formation of ferrite during cooling to the deformation temperature.

Carbon is essential for ausforming, levels of 0.3–0.4 being commonly used, although there is no reason why lower carbon contents should not be effective. Carbide forming elements are essential, partly to control the reaction kinetics but also to ensure that fine dispersions of alloy carbides are formed during treatment. The use of elements such as titanium, vanadium, niobium and molybdenum is advantageous because the carbides of these metals resist coarsening during tempering, and thus help to raise the strength of the steel.

9.4.3 Processing variables

In general, it is desirable to use as low an austenitizing temperature as is compatible with the solution of alloy carbides and the absence of excessive grain growth. The austenite grain size should be as fine as possible, not only to increase the dislocation density during deformation but also to minimize the martensite plate size on quenching from the metastable austenite bay.

Cooling from the austenitizing temperature to the metastable bay must be sufficiently rapid to avoid the formation of ferrite and, after deformation, the cooling should be fast enough to prevent the formation of bainite. The strength achieved as a result of ausforming increases as the deformation temperature is decreased, presumably because of the greater strain hardening induced in the austenite. In any case, the temperature chosen should be low enough to avoid recovery and recrystallization, but high enough to prevent bainite from forming during the deformation. Premature austenite decomposition has been found to be detrimental to mechanical properties.

The amount of deformation is a most important variable. There is a roughly linear relationship between the degree of working and the strength finally achieved, with increases between 4 and 8 MNm^{-2} per percent deformation (Fig. 9.10). One of the most significant trends is that for many steels the ductility actually increases with increasing deformation, although this only becomes significant at deformations above 30% reduction in thickness (Fig. 9.11).

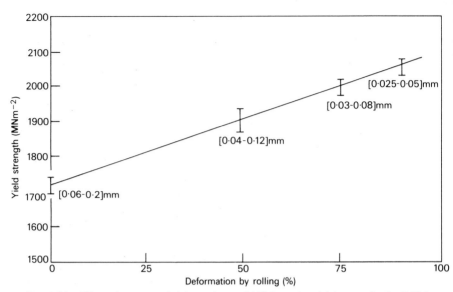

Fig. 9.10 Effect of amount of deformation at 510°C on the yield strength of a 0.32C-3.0Cr-1.5Ni-1.5Si-0.5Mo steel. A range of maternsite plate sizes (in brackets) was investigated at each deformation (Zackay, In: *National Physical Laboratory Symposium*, No. 15, HMSO, 1963)

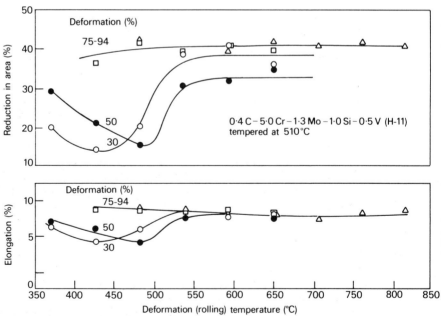

Fig. 9.11 Effect of deformation temperature on the ductility of a 0.4C-5.0Cr-1.3Mo-1.0Si-0.5V steel (H11) tempered at 510°C (Zackay, In: *National Physical Laboratory Symposium*, No. 15, HMSO, 1963)

9.4.4 Structural changes in ausforming

As might be expected, steels subjected to heavy deformation during ausforming exhibit very high dislocation densities (up to 10^{13} cm^{-2}) formed partly during deformation and partly during the shear transformation to martensite. There is evidence that a fairly uniform dislocation density is produced, in contrast to the normal dislocation cell structures obtained in heavily cold worked stable alloys. The deformation is usually carried out in the temperature range (500–600°C) in which alloy carbides would be expected to precipitate, so it is not surprising that fine alloy carbide dispersions have been detected by dark field electron microscopy. Steels most suitable for ausforming show serrated yielding during the warm deformation process which is evidence for a dynamic strain ageing process (see Chapter 2), leading to the formation of a fine dispersion of alloy carbides on the dislocation networks.

On transforming the warm worked austenite to martensite, it is likely that at least part of the dislocation substructure, together with the fine carbide dispersion, is inherited by the martensite. The martensite plate size has been shown to be very substantially smaller than in similar steels given a straight quench from the austenitizing temperature. This probably occurs because the prior dislocation arrays have been pinned by the fine precipitate and can act as barriers to martensitic plate propagation.

9.4.5 Strengthening mechanisms

Several factors must contribute to strength because any one mechanism cannot fully account for the high degree of strengthening observed. However, it seems likely that the major contributions are from the very high dislocation density and the fine dispersion of alloy carbides associated with the dislocations. It should also be added that the fine precipitate particles can act as dislocation multiplication centres during plastic deformation. The martensitic transformation is an essential part of the strengthening process, as it substantially increases the dislocation density and divides each deformed austenite grain into a large number of martensitic plates, which are much smaller than those in conventional heat treatments. It is also likely that these small plates have inherited fine dislocation substructures from the deformed metastable austenite.

The improvement in ductility is rather more difficult to explain. However, it has been observed that ausformed martensites are usually twin-free. Indeed, the presence of a high density of pinned dislocations will tend to inhibit twinning during mechanical testing. It is known that twin intersections are frequently the sites for nucleation of cracks in bcc metals, so their elimination might be expected to improve the overall ductility. Another possibility is that there is a high degree of energy of absorption during crack propagation because of the finer martensite plate size, and also because the austenite grain boundaries, favoured for crack propagation in undeformed steels, have been greatly elongated and eliminated as crack paths. Furthermore, precipitation at these boundaries should be less than in conventional treated steels because of the high concentration of carbide particles on the dislocation arrays.

9.5 Isoforming

The process of isoforming involves deformation of metastable austenite, but the deformation is continued until the transformation of austenite is complete at the deformation temperature (Fig. 9.9c). This is because the lamellar morphology of pearlite leads to low toughness in ferrite/pearlite steels, the ductile/brittle transition temperature increasing with larger volume fraction of pearlite. However, by applying deformation during the phase transformation, instead of a ferrite/pearlite aggregate, the structure produced consists of fine ferrite subgrains ($\sim 0.5\,\mu$m diameter) with spheroidized cementite particles (~ 25 nm diameter) mainly located at subgrain triple points. The misorientations across the subgrain boundaries are between $1°$ and $10°$.

As in the case of steels for ausforming, the chosen steel must have a suitable *TTT* diagram. First, it is necessary to be able to deform the austenite prior to transformation, then the transformation must be complete before deformation has ceased. Only modest increases in strength are achieved. However, there can be a very substantial improvement in toughness due to the refinement of the ferrite grain size and the replacement of lamellar cementite by spheroidized particles. However, for significant gains in toughness, deformations in excess of 70% reduction in area are needed. Finally, care must be taken to restrict deformation to temperatures at which the ferrite and pearlite reactions take place as similar deformation in the bainitic region leads to marked reductions in toughness.

9.6 High temperature thermomechanical treatments (HTMT)

In high temperature thermomechanical treatments the deformation is carried out in the stable austenite range just above Ae_3 (Fig. 9.9b), and so can be performed in steels which do not possess a suitable metastable austenite bay. The steel is then quenched to the martensitic state, followed by conventional tempering at an appropriate temperature. The strengthening achieved arises from austenite grain size refinement, typically from 10–60 μm to $\sim 3\,\mu$m, but optimum properties are often obtained if recrystallization of the austenite is avoided. As in ausforming strong carbide forming elements are beneficial, which suggests that alloy carbide precipitation occurs in the austenite during deformation. A particular advantage of this process is that optimum properties can be achieved at modest deformations ($\sim 40\%$) so that deformation can be carried out more readily on existing equipment. The HTMT process does not yield as high strengths as in ausforming but the ductility and fatigue properties are usually superior.

Clearly, HTMT is a variant of controlled rolling. However, it is normally applied to steels with higher alloying contents which can then be transformed to martensite and tempered.

9.7 Industrial steels subjected to thermomechanical treatments

Micro-alloyed steels produced by controlled rolling are a most attractive proposition in many engineering applications because of their relatively low cost, moderate strength, and very good toughness and fatigue strength, together with their ability to be readily welded. They have, to a considerable degree, eliminated quenched and tempered steels in many applications.

These steels are most frequently available in control rolled sheet, which is then coiled over a range of temperatures between 750 and 550°C. The coiling temperature has an important influence as it represents the final transformation temperature, and this influences the microstructure. The lower this temperature, under the same conditions, the higher the strength achieved. The normal range of yield strength obtained in these steels varies from about 350 to 550 MNm^{-2} (50–80 ksi). The strength is controlled both by the detailed thermomechanical treatment, by varying the manganese content from 0.5 to 1.5 wt%, and by using the microalloying additions in the range 0.03 to above 0.1 wt%. Niobium is used alone, or with vanadium, while titanium can be used in combination with the other two carbide forming elements. The interactions between these elements are complex, but in general terms niobium precipitates more readily in austenite than does vanadium as carbide or carbo-nitride, so it is relatively more effective as a grain refiner. The greater solubility of vanadium carbide in austenite underlines the superior dispersion strengthening potential of this element shared to a lesser degree with titanium. Titanium also interacts with sulphur and can have a beneficial effect on the shape of sulphide inclusions. Bearing in mind that the total effect of these elements used in conjunction is not a simple sum of their individual influence, the detailed metallurgy of these steels becomes extremely complex.

One of the most extensive applications is in pipelines for the conveyance of natural gas and oil, where the improved weldability due to the overall lower alloying content (lower hardenability) and, particularly, the lower carbon levels is a great advantage. Furthermore, as the need for larger diameter pipes has grown, steels of higher yield stress have to be used to avoid excessive wall thicknesses. In practice, wall thicknesses of 10–12.5 mm have been found to be the most convenient. Typical compositions to achieve a yield stress of around 410 MNm^{-2} (60 ksi) are:

C 0.12%   S 0.012%   Mn 1.35%   Nb 0.03%
C 0.12%   S 0.006%   Mn 1.33%   Nb 0.02%   V 0.04%

for higher yield strengths (450 MNm^{-2}):

C 0.06%   S 0.006%   Mn 1.55%   Nb 0.05%   V 0.10%

However, it should be emphasized that often higher yield stresses are achieved by control of the fabrication variables such as the temperature at which rolling is finished and the temperature used for coiling the sheet. Nitrogen is often deliberately used as an alloying element. One successful range of steels relies on vanadium to form carbo-nitride precipitates for grain size control and dispersion strengthening. In some steels, rare earth additions are made to control the inclusion shape. Typical compositions at lower and higher strength levels are given in Table 9.1.

Table 9.1 Typical compositions of micro-alloyed vanadium steels

Element	Typical composition (%)	
	Yield strength 345 (MNm^{-2}) (50 ksi)	550 (MNm^{-2}) (80 ksi)
Carbon	0.08 –0.12	0.12 –0.17
Manganese	0.75 –1.10	1.20 –1.55
Phosphorus	0.008–0.013	0.008–0.013
Sulphur	0.007–0.020	0.007–0.020
Silicon	0.05 –0.15	0.30 –0.55
Aluminium	0.03 –0.06	0.03 –0.06
Vanadium	0.03 –0.07	0.10 –0.14
Nitrogen	0.006–0.012	0.015–0.022
Cerium	0.02 –0.06	0.02 –0.06

At the higher strength levels, micro-alloyed steels are used for heavy duty truck frames, tractor components, crane booms and lighting standards, etc. The control of sulphide inclusions gives the steels a high degree of formability in cold fabrication processes. This recent development has allowed the use of HSLA steels for many applications involving substantial cold forming which previously led to cracking in the absence of rare earth additions.

Ausforming has provided some of the strongest, toughest steels so far produced, with the added advantage of very good fatigue resistance. However, they usually have high concentrations of expensive alloying elements and must be subjected to large deformations which impose heavy work loads on rolling mills. Nevertheless, these steels are particularly useful where a high strength to weight ratio is required and where cost is a secondary factor. Typical applications have included parts for undercarriages of aircraft, special springs and bolts.

The 12%Cr transformable steels respond readily to ausforming to the

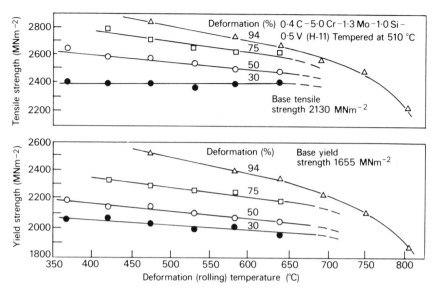

Fig. 9.12 Effect of amount and temperature of deformation on the yield and tensile strength of a 0.4C-5.0Cr-1.3Mo-1.0Si-0.5V steel (H11) (Zackay, In: *National Physical Laboratory Symposium*, No. 15, HMSO, 1963)

extent that tensile strengths of over 3000 MNm^{-2} (> 200 tsi) can be obtained in appropriate compositions. A 0.4C–6Mn–3Cr–1.5Si steel has been ausformed to a tensile strength of 3400 MNm^{-2}, with an improvement in ductility over the conventional heat treatment. Similar high strength levels with good ductility have been reported for 0.4C–5Cr–1.3Mo–1.0Si–0.5V steel (H11).(Fig. 9.12). All of these steels are sufficiently highly alloyed to allow adequate time for substantial deformation in the austenite bay of the *TTT* curve prior to transformation.

Further reading

Iron and Steel Institute, *High Strength Steels*, Special Report No. 76, 1962
Zackay, V. F. (ed), *High Strength Materials*, John Wiley, New York, 1965
Iron and Steel Institute, *Strong Tough Structural Steels*, Pubn 104, 1967
Iron and Steel Institute, *Deformation under Hot Working Conditions*, Pubn 108, 1968
The Microstructure and Design of Alloys, Third International Conference on the Strength of Metals, Cambridge UK, 1973
Microalloying 75, Proceedings of a conference held in Washington DC, October 1975
The Contribution of Physical Metallurgy to Engineering Practice, Rosenhain Centenary Conference, The Royal Society, 1976
Davenport, A. T. (ed), *Formable HSLA and Dual-phase Steels*, The Metallurgical Society of AIME, 1979

10
The embrittlement and fracture of steels

10.1 Introduction

Most groups of alloys can exhibit failure by cracking in circumstances where the apparent applied stress is well below that at which failure would normally be expected. Steels are no exception to this, and probably exhibit a wider variety of failure mechanisms than any other category of material. While ultimate failure under excessive stress must occur and can be reasonably predicted by appropriate mechanical tests, premature failure is always dangerous, involving a considerable element of unpredictability. However, a detailed knowledge of structure and of the distribution of impurities in steels is gradually leading to a much better understanding of the origins and mechanisms of the various types of cracks encountered. Furthermore, the now well-established science of fracture mechanics allows the quantitative assessment of growth of cracks in various stress situations, to an extent that it is now frequently possible to predict the stress level to which steel structures can be confidently subjected without the risk of sudden failure.

10.2 Cleavage fracture in iron and steel

Cleavage fracture is familiar in many minerals and inorganic crystalline solids as a crack propagation frequently associated with very little plastic deformation and occurring in a crystallographic fashion along planes of low indices, i.e. high atomic density. At low temperatures zinc cleaves along the basal plane, while bcc iron cleaves along $\{100\}$ planes (Fig. 10.1), as do all bcc metals. This behaviour would appear to be an intrinsic characteristic of iron but it has been shown that iron, highly purified by zone refining and containing minimal concentrations of carbon, oxygen and nitrogen, is very ductile even at extremely low temperatures. For example, at 4.2 K reductions in area in tensile tests of up to 90 % have been observed with iron specimens of the highest available purity. As the carbon and nitrogen content of the iron is increased, the transition from ductile to brittle cleavage behaviour takes place at increasing temperatures, until in some steels this can occur at ambient and above-ambient temperatures. Clearly, the significant variables in such a transition are of great basic and practical importance.

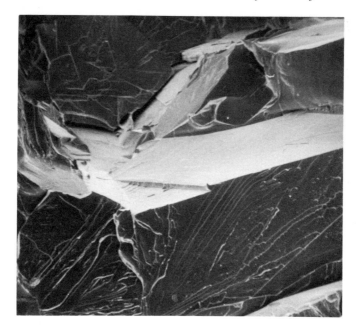

Fig. 10.1 Cleavage fracture in pure iron-0.04P at $-55°$C (Shell) ×60

The propagation of a cleavage crack in iron and steel requires much less energy than that associated with the growth of a ductile crack. This is easily shown by carrying out impact tests in a pendulum apparatus (Charpy, Izod and Hounsfield type tests) over a range of temperature. The energy absorbed by the specimen from the pendulum when plotted as a function of temperature usually exhibits a sharp change in slope (Fig. 10.2) as the mode of fracture changes from ductile to brittle. These impact transition curves are a simple way of defining the effect of metallurgical variables, e.g. heat treatment (Fig. 10.2) on the fracture behaviour of a steel from which a fairly precise transition temperature, T_c, can be readily obtained for a particular heat treatment. However, it should be emphasized that T_c is not an absolute value and it is likely to change appreciably as the mode of testing is altered. It nevertheless provides a simple way of comparing the effects of metallurgical variables on the fracture behaviour.

In recent years, more sophisticated tests have been developed in which it is recognized that the propagation of the fracture is the important stage. These fracture toughness tests use notched and pre-cracked specimens, the cracks being initiated by fatigue. The stress intensity factor, K, at the root of the crack is defined in terms of the applied stress σ and the crack size c:

$$K = \sigma(\pi c)^{\frac{1}{2}}$$

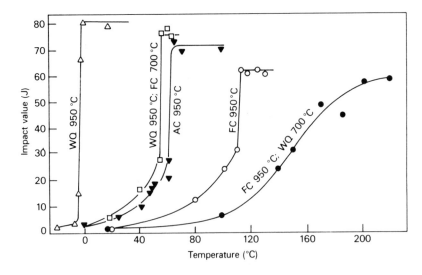

Fig. 10.2 Effect of heat treatments on the impact transition temperature of a pure iron-0.12C alloy (after Allen et al., JISI, 1953, **174**, 108)

When a critical stress intensity factor K_C is reached, the transition to rapid fracture takes place.

10.3 Factors influencing the onset of cleavage fracture

There are several factors, some interrelated, which play an important part in the initiation of cleavage fracture:
(1) The temperature dependence of the yield stress
(2) The development of a sharp yield point
(3) Nucleation of cracks at twins
(4) Nucleation of cracks at carbide particles
(5) Grain size.

All body-centred cubic metals including iron show a marked temperature dependence of the yield stress, even when the interstitial impurity content is very low, i.e. the stress necessary to move dislocations, the Peierls-Nabarro stress, is strongly temperature dependent. This means that as the temperature is lowered the first dislocations to move will do so more rapidly as the velocity is proportional to the stress, and so the chances of forming a crack nucleus, e.g. by dislocation coalescence (Fig. 10.3), will increase. Fig. 10.3 shows schematically two ways in which dislocation pile-ups could nucleate cracks.

The interstitial atoms, carbon and nitrogen, will cause the steel to exhibit a sharp yield point (Chapter 2) either by the catastrophic break-away of dislocations from their interstitial atom atmospheres (Cottrell-Bilby

The embrittlement and fracture of steels 189

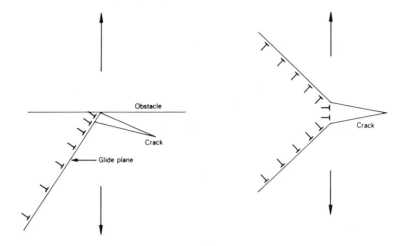

Fig. 10.3 Schematic diagram of dislocation mechanisms for crack nucleation

theory), or by the rapid movement of freshly generated dislocations (Gilman-Johnson theory). In either case, the conditions are suitable for the localized rapid movement of dislocations as a result of high stresses which provides a favourable situation for the nucleation of cracks by dislocation coalescence.

The flow stress of iron increases rapidly with decreasing temperature (Fig. 2.2) to a point where the critical stress for deformation twinning is reached, so that this becomes a significant deformation mechanism. It has been shown that cracks are preferentially nucleated at various twin configurations, e.g. at twin intersections and at points where twins contact grain boundaries, so that, under the same conditions, crack propagation is more likely in twinned iron. It should also be noted that the temperature dependence of the flow stress makes plastic deformation more difficult at the tip of a moving crack, so less plastic blunting of the crack tip will take place at low temperatures, thus aiding propagation.

So far, we have discussed crack nucleation mechanisms which can take place in single phase material, e.g. relatively pure iron, but in the presence of a second phase such as cementite it is still easier to nucleate cracks. Plastic deformation can crack grain boundary cementite particles or cementite lamellae in pearlite so as to produce micro-cracks (Fig. 10.4) which, in certain circumstances, propagate to cause catastrophic cleavage failure (Fig. 10.5). Recent work supports the view that this microstructural parameter is extremely important in determining the fracture characteristics of a steel. Brittle inclusions such as alumina particles or various silicates found in steels can also be a source of crack nuclei.

190 Steels—Microstructure and Properties

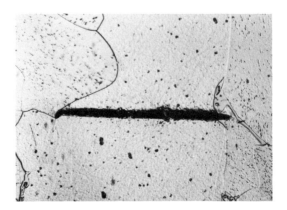

Fig. 10.4 Nucleation of a cleavage crack at a carbide particle in a low carbon steel (Knott). Optical micrograph, ×400

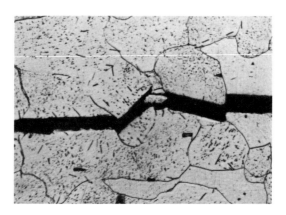

Fig. 10.5 Transgranular propagation of a crack in a low carbon steel (Knott). Optical micrograph, ×275

Grain size is a particularly important variable for, as the ferrite grain size is reduced, the transition temperature T_c is lowered, despite the fact that the yield strength increases. This is, therefore, an important strengthening mechanism which actually improves the ductility of the steel. It has been shown by Petch that T_c is linearly related to $\ln d^{-\frac{1}{2}}$, and an appropriate relationship of this type can be derived from a dislocation model involving the formation of crack nuclei at dislocation pile-ups at grain boundaries. The smaller the grain size, the smaller the number of dislocations piling-up where a slip band arrives at a boundary. Bearing in mind that the shear stress at the head of such a pile-up is $n\tau$ where n is the number of dislocations and τ is the shear stress in the slip direction, it follows that as the grain size is reduced, n will be smaller and the local stress concentrations at grain boundaries will be correspondingly less. This situation will lead to less crack nuclei regardless of whether they are formed by dislocation coalescence or by dislocation pile-ups causing carbides to crack or by twinning interactions.

10.4 Criterion for the ductile/brittle transition

The starting point of all theories on brittle fracture is the work of Griffith, who considered the condition needed for propagation of a pre-existing crack, of length $2c$, in a brittle solid. When the applied stress σ is high enough, the crack will propagate and release elastic energy. This energy U_e in the case of thin plates (plane stress) is:

$$U_e = -\frac{\pi c^2 \sigma^2}{E} \text{ per unit plate thickness} \quad (10.1)$$

where E = Young's modulus. The term is negative because this energy is released. However as the crack creates two new surfaces, each with energy $= 2c\gamma$, there is a positive surface energy term U_s:

$U_s = 4c\gamma$ where γ = surface energy per unit area.

Griffith showed that the crack would propagate if the increase in surface energy, U_s, was less than the decrease in elastic energy U_e. The equilibrium position is defined as that in which the change in energy with crack length is zero:

$$\frac{dU}{dc} = \frac{d(U_e + U_s)}{dc} = 0 \quad (10.2)$$

This is the elastic strain energy release rate, usually referred to as G.

$$\therefore \quad \left(-\frac{2\pi c \sigma^2}{E} + 4\gamma \right) = 0$$

and
$$\sigma_f = \left(\frac{2\gamma E}{\pi c} \right)^{\frac{1}{2}} \quad (10.3)$$

where σ_f is the fracture stress, which is defined as that just above which energy is released and the crack propagates. This equation shows that the stress σ is inversely related to crack length, so that as the crack propagates the stress needed drops and the crack thus accelerates. Orowan pointed out that in crystalline solids plastic deformation will occur both during nucleation of the crack, and then at the root of the crack during propagation. This root deformation blunts the crack and, in practice, means that more energy is needed to continue the crack propagation. Thus the

Griffith equation is modified to include a plastic work term γ_p:

$$\sigma_f = \left(\frac{E(2\gamma + \gamma_p)}{\pi c}\right)^{\frac{1}{2}} \tag{10.4}$$

It has been found that $\gamma_p \gg \gamma$, hence the condition for crack spreading in a crystalline solid such as iron is

$$\sigma_f = \left(\frac{E\gamma_p}{\pi c}\right)^{\frac{1}{2}} \tag{10.5}$$

The local stress field at the crack tip is usually characterized by a parameter K, the stress intensity factor, which reaches a critical value K_C when propagation takes place. This critical value is given by

$$K_C = \sigma_f \sqrt{\pi c} \tag{10.6}$$

In plane stress conditions

$$K_C = \sqrt{EG_C}$$

where G_C = the critical release rate of strain energy.

In plane strain conditions, the critical value of strain energy release rate is $G_{1C} = \gamma_p$ where

$$\sigma_f = \left(\frac{EG_{1C}}{\pi(1-v^2)c}\right)^{\frac{1}{2}} \tag{10.7}$$

where v = Poisson's ratio

The critical value of stress intensity, K_{1C}, is then related to G_{1C}:

$$K_{1C} = \left(\frac{EG_{1C}}{(1-v^2)}\right)^{\frac{1}{2}} \tag{10.8}$$

The fracture toughness of a steel is often expressed as a K_{1C} value obtained from tests on notched specimens which are precracked by fatigue, and are stressed to fracture in bending or tension.

The nucleation and the propagation of a cleavage crack must be distinguished clearly. Nucleation occurs when a critical value of the *effective shear stress* is reached, corresponding to a critical grouping, ideally a pile-up, of dislocations which can create a crack nucleus, e.g. by fracturing a carbide particle. In contrast, propagation of a crack depends on the magnitude of the local *tensile stress* which must reach a critical level σ_f. Simple models of slip-nucleated fracture assume either interaction of dislocations or cracks formed in grain boundary carbides. However, recently it has been realized that both these structural features must be taken into account in deriving an expression for the critical fracture stress σ_f. This critical stress does not appear to be temperature dependent. At low temperatures the yield stress is higher, so the crack propagates when the plastic zone ahead of the crack is small, whereas at higher temperatures, the yield stress being smaller, a larger plastic zone is required to achieve the critical local tensile stress σ_f.

This tensile stress σ_f has been determined for a wide variety of mild steels, and has been shown to vary roughly linearly with $d^{-\frac{1}{2}}$ (Fig. 10.6). The scatter probably arises from differences in test temperature and carbide dimensions. This is conclusive evidence for the role of finer grain sizes in increasing the resistance to crack propagation. Regarding grain boundary carbide size, effective crack nuclei will occur in particles above a certain critical size so that, if the size distribution of carbide particles in a particular steel is known, it should be possible to predict its critical fracture stress. Therefore, in mild steels in which the structure is essentially ferrite grains containing carbide particles, the particle size distribution of carbides is the most important factor. In contrast, in bainitic and martensitic steels the austenite grains transform to lath structures where the lath width is usually between 0.2 and 2 μm. The laths occur in bundles or packets (see Chapter 5) with low angle boundaries between the laths. Larger misorientations occur across packet boundaries. In such structures, the packet width is the main microstructural feature controlling cleavage crack propagation.

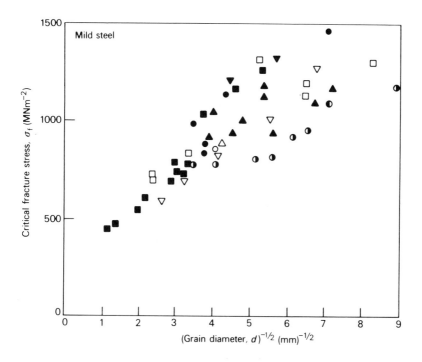

Fig. 10.6 Dependence of local fracture stress σ_f on the grain size of mild steel. Date from many sources (Knott)

The critical local fracture stress σ_f has been related to the two types of structure, as follows.

For ferritic steels with spheroidal carbide particles:

$$\sigma_f = \left(\frac{\pi E \gamma_p}{2C_o}\right)^{\frac{1}{2}} \qquad (10.9)$$

where C_o = carbide diameter.

For bainitic and martensitic steels with packets of laths:

$$\sigma_f = \left(\frac{4E\gamma_p}{(1-v^2)d_p}\right)^{\frac{1}{2}} \qquad (10.10)$$

where d_p = packet width and v = Poisson's ratio.

10.5 Practical aspects of brittle fracture

At the onset of fracture, elastic energy stored in the stressed steel is only partly used for creation of the new surfaces and the associated plastic deformation and the remainder provides kinetic energy to the crack. Using a Griffth-type model, the crack velocity v can be shown to be:

$$v = \sqrt{\frac{2\pi}{k}} \sqrt{\frac{E}{\rho}} \left(1 - \frac{c_0}{c}\right)^{\frac{1}{2}}$$

where c_0 = critical size, c = half crack size at a given instant, ρ = density and k = constant. This relation shows that the velocity increases with increasing crack size and reaches a limiting value $v_{\lim}$ at large values of c. In practice, $v_{\lim}$ is between 0.4 and 0.5 of the speed of sound, so brittle fracture occurs with catastrophic rapidity, as many disasters testify.

The phenomenon of brittle fracture became particularly prevalent with the introduction of welding as the major steel fabrication technique. Previously, brittle cracks often stopped at the joints of riveted plates but the steel structures resulting from welding provided continuous paths for crack propagation. Added to this, incorrect welding procedures can give rise to high stress concentrations and also to the formation of weld-zone cracks which may initiate brittle fracture. While brittle failures of steels have been experienced since the latter half of the nineteenth century when steel began to be used widely for structural work, the most serious failures have occurred in more recent years as the demand for integral large steel structures has greatly increased, e.g. in ships, pipelines, bridges and pressure vessels. Spectacular failures took place in many of the all-welded Liberty ships produced during the Second World War, when nearly 1500 incidents involving serious brittle failure were recognized and nineteen ships broke completely in two without warning. Despite our increasing understanding of the phenomenon and the great improvements in steel making and in welding since then, serious brittle failures still occur (Fig. 10.7), a constant

Fig. 10.7 Brittle fracture of a thick-walled steel pressure vessel (The Welding Institute)

reminder that human error and lack of scientific control can be disastrous.

Bearing in mind the temperature dependence of the failure behaviour, and the widening use of steels at low temperatures, e.g. in Arctic pipelines, for storage of liquid gases etc., it is increasingly necessary to have steels with very low transition temperatures and high fracture toughness. While there are many variables to consider in achieving this end, including the detailed steel-making practice, the composition including trace elements and the fabrication processes involved, the most important is probably grain size refinement. The development of high strength low alloy steels (HSLA) or micro-alloyed steels (Chapter 9), in the manufacture of which controlled rolling plays a vital part, has led to the production of structural steels with grain sizes often less than 10 μm combined with good strength levels (yield strength between 400 and 600 MNm^{-2}) and low transition temperatures. In these steels, to which small concentrations ($< 0.1\%$) of niobium, vanadium or titanium are added, the carbon level is usually less than 0.15 % and often below 0.10 %, so that the carbide phase occupies a small volume fraction. In any case, cementite, which forms relatively coarse particles or lamellae in pearlite, is partly replaced by much finer dispersions of alloy carbides, NbC etc. Addition of certain other alloying elements to steel, notably manganese and nickel, results in a lowering of the transition temperature. For example, alloy steels with 9 % nickel and less than 0.1 % carbon have a sufficiently low transition temperature to be able to be used for large containers of liquid gases, where the temperature can be as low as 77K. Below this temperature, austenitic steels have to be used. Of the elements unavoidably present in steels, phosphorus, which is substantially soluble in α-iron, raises the transition temperature and thus must be kept to as low a concentration as possible. On the other hand, sulphur has a very low solubility, and is usually present as manganese sulphide with little effect on the transition temperature but with an important role in ductile fracture. Oxygen is an embrittling element even when present in very small concentrations. However, it is easily removed by deoxidation practice involving elements such as manganese, silicon and aluminium.

Finally, the fabrication process is often of crucial importance. In welding it is essential to have a steel with a low carbon equivalent, i.e. a factor incorporating the effects on hardenability of the common alloying elements. A simple empirical relationship, as a rough guide, is:

$$\% \text{ carbon equivalent} = C + Mn + \left(\frac{Cr + Mo + V}{5}\right) + \left(\frac{Ni + Cu}{15}\right) \text{ (in wt\%)}$$

where a steel with an equivalent of less than 0.45 should be weldable with modern techniques. The main hazard in welding in is the formation of martensite in the heat affected zone (HAZ), near the weld, which can readily lead to microcracks. This can be avoided, not only by control of hardenability but also by preheating the weld area to lead to slower cooling after welding or by post heat treatment of the weld region. However, in some high strength steels, slower cooling may result in the formation of upper bainite in the HAZ which encourages cleavage fracture.

Attention must also be paid to the possibility of hydrogen absorption leading to embrittlement. The presence of hydrogen in steels often leads to disasterous brittle fracture, e.g. there have been many failures of high strength steels into which hydrogen was introduced during electroplating of protective surface layers. Concentrations of a few parts per million are often sufficient to cause failure. While much hydrogen escapes from steel in the molecular form during treatment, some can remain and precipitate at internal surfaces such as inclusion/matrix and carbide/matrix interfaces, where it forms voids or cracks. Cleavage crack growth then occurs slowly under internal hydrogen pressure, until the critical length for instability is reached, and failure occurs rapidly. Hydrogen embrittlement is not sensitive to composition, but to the strength level of the steel, the problem being most pronounced in high strength alloy steels. It is frequently encountered after welding (Fig. 10.8), where it can be introduced by use of damp welding electrodes, leading to cracking which is variously referred to as *underbead cracking*, *cold cracking* and *delayed cracking*. This phenomenon can be minimized by the use of welding electrodes with very low hydrogen contents, which are oven-dried prior to use.

10.6 Ductile or fibrous fracture

10.6.1 General

The higher temperature side of the ductile/brittle transition is associated with a much tougher mode of failure, which absorbs much more energy in

The embrittlement and fracture of steels 197

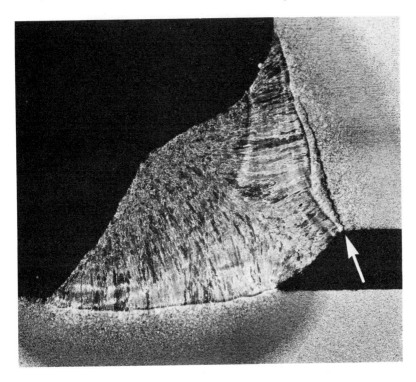

Fig. 10.8 Cleavage crack due to hydrogen embrittlement in the heat affected zone of a weld in BS 968 steel (The Welding Institute)

the impact test. While the failure mode is often referred to as ductile fracture, it could be described as rupture, a slow separation process which, although transgranular, is not markedly crystallographic in nature. Scanning electron micrographs of the ductile fracture surface (Fig. 10.9), in striking contrast to those from the smooth faceted cleavage surface, reveal a heavily dimpled surface, each depression being associated with a hard particle, either a carbide or non-metallic inclusion.

It is now well-established that ductile failure is initiated by the nucleation of voids at second phase particles. In steels these particles are either carbides, sulphide or silicate inclusions. The voids form either by cracking of the particles, or by decohesion at the particle/matrix interfaces, so it is clear that the volume fractions, distribution and morphology of both carbides and of inclusions are important in determining the ductile behaviour, not only in the simple tensile test, but in complex working operations. Therefore, significant variables, which determine ductility of steels, are to be found in the steel-making process, where the nature and

198   *Steels—Microstructure and Properties*

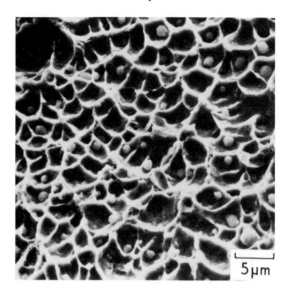

Fig. 10.9 Ductile fracture of a low alloy steel (R. F. Smith). Stereoscan

distribution of inclusions is partly determined, and in subsequent solidification and working processes. Likewise, the carbide distribution will depend on composition and on steel-making practice, and particularly on the final heat treatment involving the transformation from austenite, which largely determines the carbide size, shape and distribution.

The formation of voids begins very early in a tensile test, as a result of high stresses imposed by dislocation arrays on individual hard particles. Depending on the strength of the particle/matrix bond, the voids occur at varying strains, but for inclusions in steels the bonding is usually weak so voids are observed at low plastic strains. These elongate under the influence of the tensile stress but, additionally, a lateral stress is needed for them to grow sideways and link up with adjacent voids forming *micronecks*. These necks progressively part (Fig. 10.10) leading to the ductile fracture surfaces with a highly dimpled appearance. The second phase particles (MnS) can be clearly seen in Fig. 10.10.

Many higher strength steels exhibit lower work hardening capacity as shown by relatively flat stress-strain curves in tension. As a result, at high strains the flow localizes in shear bands, where intense deformation leads to decohesion, a type of shear fracture. While the detailed mechanism of this process is not yet clear, it involves the localized interaction of high dislocation densities with carbide particles.

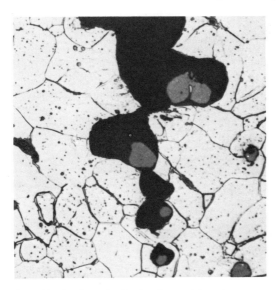

Fig. 10.10 Growth of a ductile crack in a free-cutting mild steel containing sulphides (R. F. Smith). Optical micrograph, ×600

10.6.2 Role of inclusions in ductility

It is now generally recognized that the deformability of inclusions is a crucial factor which plays a major role, not only in service where risk of fracture exists, but also during hot and cold working operations such as rolling, forging, machining. Kiessling has divided the inclusions found in steels into five categories relating to their deformation behaviour.
(1) Al_2O_3 and calcium aluminates – these arise during deoxidation of molten steels. They are brittle solids, which are in practical terms undeformable at all temperatures.
(2) Spinel type oxides $AO\text{-}B_2O_3$ – these are undeformable in the range RT to 1200° C, but may be deformed above this temperature.
(3) Silicates of calcium, manganese, iron and aluminium in various proportions – these inclusions are brittle at room temperature, but increasingly deformable at higher temperatures. The formability increases with decreasing melting point of the silicate, e.g. from aluminium silicate to iron and manganese silicates.
(4) FeO, MnO and (FeMn)O – these are plastic at room temperature, but appear gradually to become less plastic above 400° C.
(5) Manganese sulphide MnS – this common inclusion type is deformable, becoming increasingly so as the temperature falls. There are three main

types of MnS inclusion dependent on their mode of formation, which markedly influences their morphology:

Type I: globular, formed only when oxygen is present in the melt, e.g. in rimming steels (Fig. 10.11a)
Type II: interdendritic eutectic form, familiar in killed steels (Fig. 10.11b)
Type III: random angular particles, found in fully deoxidised steels (Fig. 10.11c).

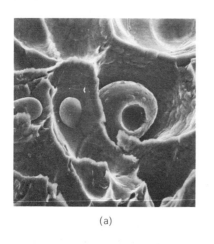

(a)

(b)

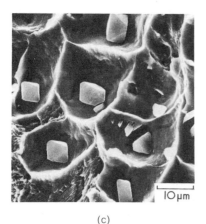

(c)

Fig. 10.11 Manganese sulphide inclusions in steels: a, Type I; b, Type II; c, Type III (T. J. Baker). Stereoscans, ×1000

It is now known that ductile failure can be associated with any of the types of inclusion listed above, from the brittle alumina type to the much more ductile sulphide inclusions. However, the inclusions are more effective in initiating ductile cracks above a critical size range. The coarser particles lead to higher local stress concentrations, which cause localized rupture and microcrack formation. Some quantitative work has now been done on model systems, e.g. iron-alumina where the progressive effect on ductility of increasing volume fraction of alumina is readily shown. The reduction in yield stress, also observed, arises from stress concentrations around the inclusions and is already evident at relatively low volume fractions.

The presence of particles in the size range $1-35\,\mu$m broadens substantially the temperature range of the ductile/brittle transition in impact tests and also lowers the energy absorbed during ductile failure, the *shelf energy*. A fine dispersion of non-brittle type inclusions can delay cleavage fracture by localized relaxation of stresses with a concomitant increase in yield stress.

Regarding cyclic stressing, it appears that inclusions must reach a critical size before they can nucleate a fatigue crack but the size effect depends also very much on the particle shape, e.g. whether spherical or angular. It has been found in some steels, e.g. ball bearing steels, that fatigue cracks originate only at brittle oxide inclusions, and not at manganese sulphide particles or oxides coated with manganese sulphide. In such circumstances the stresses which develop at particle interfaces with the steel matrix, as a result of differences in thermal expansion, appear to play an important part. It has been found that the highest stresses arise in calcium aluminates, alumina and spinel inclusions, which have substantially smaller thermal expansion coefficients than steel. These inclusions have the most deleterious effects on fatigue life.

The behaviour of ductile inclusions such as MnS during fabrication processes involving deformation has a marked effect on the ductility of the final product. Types I and III manganese sulphide will be deformed to ellipsoidal shapes, while Type II colonies will rotate during rolling into the rolling plane, giving rise to very much reduced toughness and ductility in the transverse direction. This type of sulphide precipitate is the most harmful so efforts are now made to eliminate it by addition of strong sulphide forming elements such as Ti, Zr and Ca. The lack of ductility is undoubtedly encouraged by the formation at the inclusion interfaces of voids because the MnS contracts more than the iron matrix on cooling, and the interfacial bond is probably insufficiently strong to suppress void formation. The variation in ductility with direction in rolled steels can be extreme because of the directionality of the strings of sulphide inclusions, and this in turn can adversely affect ductility during many working operations.

Cracking can also occur during welding of steel sheet with low transverse ductility. This takes place particularly in the parent plate under butt welds, the cracks following the line of the sulphide inclusion stringers. The phenomenon is referred to as *lamellar tearing* (Fig. 10.12).

10.6.3 Role of carbides in ductility

The ductility of steel is also influenced by the carbide distribution which can

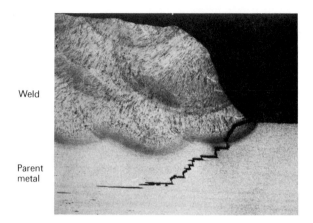

Fig. 10.12 Lamellar tearing near a weld (The Welding Institute)

vary from spheroidal particles to lamellar pearlitic cementite. Comparing spheroidal cementite with sulphides of similar morphology, the carbide particles are stronger and do not crack or exhibit decohesion at small strains, with the result that a spheroidized steel can withstand substantial deformation before voids are nucleated and so exhibits good ductility. The strain needed for void nucleation decreases with increasing volume fraction of carbide and so can be linked to the carbon content of the steel.

Pearlitic cementite also does not crack at small strains, but the critical strain for void nucleation is lower than for spheroidized carbides. Another factor which reduces the overall ductility of pearlitic steels is the fact that once a single lamella cracks, the crack is transmitted over much of a pearlite colony leading to well-defined cracks in the pearlite regions. The result is that the normal ductile dimpled fractures are obtained with fractured pearlite at the base of the dimples.

The effects of second phases on the ductility of steel are summarized in Fig. 10.13, where the sulphides are shown to have a more pronounced effect than either carbide distribution. This arises because, in the case of the sulphide inclusions, voids nucleate at a very early stage of the deformation process. The secondary effect of the particle shape both for carbides and sulphides is also indicated.

10.7 Intergranular embrittlement

While cleavage fracture in steels is a common form of embrittlement, in many cases the embrittlement is intergranular (IG), i.e. it takes place along the grain boundaries, usually the former austenitic boundaries (Fig. 10.14). This behaviour is encountered in as-quenched steels, on tempering (*temper*

The embrittlement and fracture of steels   203

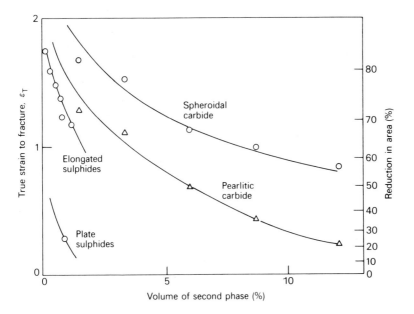

Fig. 10.13 Effect of second phase particles on the ductility of steel (Gladman et al., In: *Effect of Second-phase Particles on the Mechanical Properties of Steel*, Iron and Steel Institute, 1971)

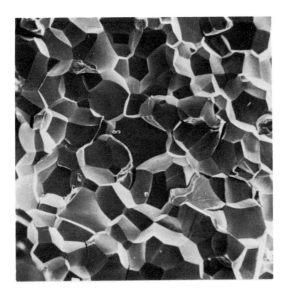

Fig. 10.14 Intergranular embrittlement of an Fe-0.26P alloy after holding at 500°C (Shell). ×80

embrittlement), after heating at very high austenitizing temperatures (*overheating* and *burning*), and in *rock-candy fracture* in cast steels. These forms of embrittlement are exhibited at or around room temperature. There are, however, other phenomena involving failure along grain boundaries which are essentially high temperature events, e.g. hot-shortness during the hot working of steels and high temperature creep failure. It is clear that no one mechanism will explain the various types of embrittlement, but the processes leading to IG fracture all lead to reduced cohesion along the grain boundaries. This can arise in different ways but the most relevant appear to be

(1) Segregation of solute atoms preferentially to grain boundaries
(2) Distribution of second phase particles at grain boundaries.

These phenomena reduce the work of fracture, i.e. the $(2\gamma + \gamma_p)$ term in (Equation 10.4) either by lowering the grain boundary energy by segregation, or by reducing the plastic work term γ_p by having particles which more easily provide crack nuclei.

10.7.1 Temper embrittlement

Many alloy steels when tempered in the range 500°C–650°C following quenching to form martensite become progressively embrittled in an intergranular way. A similar phenomenon can also occur when the steels are continuously cooled through the critical range. It is revealed by the effect on the notched bar impact test, where the transition temperature is raised and the shelf energy lowered, the transgranular fracture mode being replaced by an IG mode below the transition temperature (Fig. 10.15).

Fig. 10.15 Temper embrittlement of a 4.5Ni-1.5Cr-0.3C steel fractured at 77 K (Knott). ×800

This phenomenon is now known to be associated with the segregation of certain elements to the grain boundaries, which reduce the intergranular cohesion of iron. Elements which segregate fall into three groups of the Periodic Classification (Table 10.1). It has been shown that many of these elements reduce the surface energy of iron substantially and would, therefore, be expected, to lower the grain boundary energy and to reduce cohesion. Moreover, the actual segregation of atoms to the boundaries has been conclusively demonstrated by Auger electron spectroscopy on specimens fractured intergranularly within the vacuum system of the apparatus.

Table 10.1 Elements which segregate to iron grain boundaries

Group	IVB	VB	VIB
	C	N	O
	Si	P	S
	Ge	As	Se
	Sn	Sb	Te
		Bi	

This technique has allowed the precise determination of the concentrations of segregating species at the boundaries, usually expressed in terms of fractions of a monolayer of atoms. These fractions vary between about 0.3 and 2.0 for steels containing the above elements, usually in bulk concentrations well below 0.1 wt%.

With the individual elements, the tendency to embrittle appears to increase both with Group and Period number, i.e. S, Se and Te in increasing order are the most surface active elements in iron. However, it is doubtful whether they contribute greatly to temper embrittlement because they combine strongly with elements such as Mn and Cr which effectively reduce their solubility in iron to very low levels. While the elements in Groups IVB and VB are less surface active, they play a greater role in embrittlement because they interact with certain metallic elements, e.g. Ni and Mn, which are common alloying elements in steels. These interactions lead to cosegregation of alloy element and impurity elements at the grain boundaries, and to resultant lowering of cohesion by the impurity element. Analysis of the composition of grain boundaries by Auger spectroscopy has confirmed strong interactions between Ni–Sb, Ni–P, Ni–Sn and Mn–Sb. Fig. 10.16 shows the grain boundary concentrations for three of these interactions in Ni–Cr steels, while the relative effects of Sb, Sn and P on the transition temperature of Ni–Cr steels are shown in Fig. 10.17.

Therefore, the driving force for cosegregation to boundaries is a stronger interaction between the alloying element and the impurity element than between either of these and iron. If the interaction is too strong, segregation does not take place. Instead a scavenging effect is obtained, as exemplified

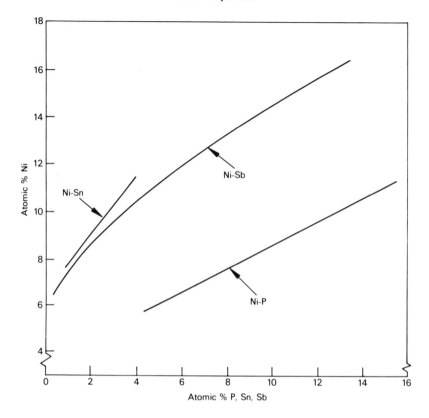

Fig. 10.16 Interrelation between concentrations of Sn, Sb and P and of Ni at grain boundaries in Ni-Cr steels of constant hardness and grain size (McMahon, *Materials Science and Engineering*, 1976, **25**, 233)

by Ti–P and Mo–P interactions in Ni–Cr steels. In this connection it is well known that molybdenum additions to Ni–Cr steels can eliminate temper embrittlement. A third inter-alloy effect is also possible which is that one alloying element, e.g. Cr, promotes the segregation of Ni and P, also Ni and Sb.

In addition to solute atom segregation to boundaries, there are also microstructural factors which influence the intensity of temper embrittlement. In most alloy steels in which this phenomenon is encountered the grain boundaries are also the sites for carbide precipitation, either cementite or alloy carbides. It is likely that these provide the sites for IG crack nuclei. As in the nucleation of cleavage fracture, dislocations impinge on a grain boundary carbide particle and as it is not deformable the carbide will either crack or the ferrite/carbide interface will part. The latter separation is more likely if the interfacial energy has been reduced by segregation of impurity atoms to it. This can occur by rejection of these impurity atoms during the

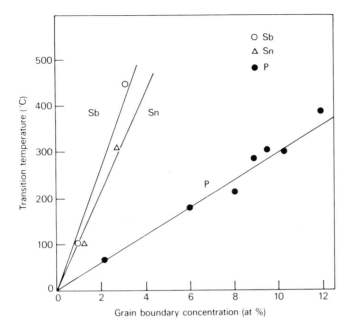

Fig.10.17 Effect of grain boundary concentrations of P, Sb and Sn on the ductility of Ni-Cr steels of constant hardness and grain size (McMahon, *Materials Science and Engineering*, 1976, **25**, 233)

growth of the carbide or by equilibrium segregation. Interfacial separation has been observed in iron containing coarse grain boundary iron carbide, the interfaces of which contained Sb, As, Sn or P. The effectiveness of this nucleating stage of IG crack formation will be influenced by the extent of grain boundary carbide and the concentration of surface active impurities in the steel, in particular at carbide/matrix interfaces.

The propagation of the grain boundary crack will depend not only on the cohesion of the boundary but also on the relative toughness of the grain interior. For example, if the grain interior has a microstructure which gives high toughness, the IG crack nucleus is more likely to propagate along the boundary. Further, as the yield stress of a steel rises sharply with decreasing temperature IG failure will, like cleavage fracture, be encouraged by reducing the testing temperature. Increasing the austenite grain size, by use of high austenitizing temperatures, under the same conditions, should increase the embrittlement because the size of the dislocation arrays impinging on the grain boundary carbides will be larger and thus more effective in forming crack nuclei.

The optimum temperature range for temper embrittlement is between 500 and 575°C. However, in some steels embrittlement occurs in the range 250–400°C. This phenomenon is called 350° (500°F) embrittlement, and occurs at too low a temperature to attribute it to the diffusion of metalloids such as Sb to the austenite grain boundaries. It seems more likely that it could arise from smaller and more mobile atoms, e.g. P, which would be rejected during grain boundary growth of iron carbide which takes place in this temperature range. However, the morphology of the grain boundary Fe_3C, if predominantly sheet-like, could be a prime cause of low ductility in this temperature range.

Stress corrosion cracking involves failure by cracking in the presence of both a stress and of a corrosive medium. It can occur in either a transgranular or an intergranular mode. The latter mode appears to be encouraged in some alloy steels by heat treatments which produce temper embrittlement. For example, a temper embrittled Cr-Mo steel cracks along the grain boundaries when stressed in a boiling NaOH solution. Use of a heat treatment to remove the temper embrittlement also removes the sensitivity to stress corrosion.

10.7.2 Overheating and burning

Many alloy steels when held in the range 1200–1400°C and subsequently heat treated by quenching and tempering, fail intergranularly along the original austenitic boundaries. There is strong evidence to suggest that this phenomenon is associated with the segregation of sulphur to the austenite grain boundaries at the high temperature, and indeed the phenomenon is not obtained when the sulphur content of a steel is less than 0.002%. Sulphur has been shown to be one of the most surface active elements in iron. Work by Goux and colleagues on pure iron-sulphur alloys has shown that an increase in sulphur content from 5 to 25 ppm raises the ductile/brittle transition temperature by over 200°C. Further, Auger spectroscopy on the intergranular fracture surfaces has given direct evidence of sulphur segregation. However, this embrittling effect of sulphur as a result of equilibrium segregation is only seen in pure iron and not in steels where there are other impurity elements, and also where interaction of sulphur occurs with alloying elements, notably manganese and chromium.

The presence of manganese substantially lowers the solubility of sulphur in both γ- and α-iron, with the result that when sulphur segregates to high temperature austenite boundaries, manganese sulphide is either formed there or during subsequent cooling. In either case, the manganese sulphide particles lying on the austenite boundaries are revealed by electron microscopy of the intergranular fracture surfaces where they are associated with small dimples. Typically the MnS particles are about 0.5 μm while the dimples are approximately 2–5 μm in diameter. Thus, the grain boundary fracture process is nucleated by the sulphide particles, and the mode of fracture will clearly be determined by the size distribution, which will in turn be controlled by the rate of cooling from the austenite temperature, assuming that MnS forms during cooling. With very slow cooling rates, the intergranular fracture is replaced by cleavage or transgranular fibrous fracture as the grain boundary sulphide distribution is too coarse. Oil quenching from the austenitizing temperature does not eliminate the phenomenon which is accentuated on tempering in the range 600–650°C. This arises from the redistribution of carbides which will strengthen the grain interiors, and by precipitation at the grain boundaries which may further reduce grain boundary ductility.

When very high austenitizing temperatures are used (1400–1450°C) extensive MnS precipitate is formed, often in impressive dendritic forms (Fig. 10.18). In extreme cases, partial formation of liquid phase occurs (liquation) which, on subsequent heat treatment, greatly accentuates the intergranular embrittlement. In the absence of manganese, e.g. in wrought iron, liquid films of the iron-iron sulphide eutectic cause embrittlement during hot working processes down to 1000°C (*hot shortness*). The fact that in normal steels burning occurs only at very high temperatures should not be allowed to detract from its significance. The phenomenon may well intrude in high temperature working processes such as forging if temperature control is not exact, but in any case it can certainly be significant in steels which are cast, and by definition pass through the burning and overheating temperature range. In many cases intergranular fracture is encountered in cast alloy steels where the as-cast grain structure is clearly involved. Examination of the fractures reveals extensive grain boundary sheets of manganese sulphide, often only 0.2–0.5 μm thick but covering large areas. Marked embrittlement can occur in the as-cast state or after subsequent heat treatment in the range 500–650°C, and is often referred to as cast brittleness or rock candy fracture. Precipitation of aluminium nitride may also play an important role in this type of fracture.

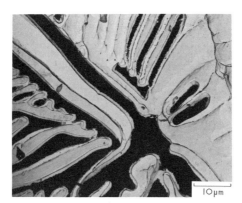

Fig. 10.18 Fe-1Mn-0.4C-0.02S air cooled from 1445°C then fractured (Brammar). Extraction replica, ×1000

Further reading

Averbach, B. L. et al. (eds), *Fracture 1959*, John Wiley, 1959
Low, J. R., The fracture of metals. *Progress in Materials Science*, Pergamon Press, **12**, (1), 1963
Tetelman, A. S. and McEvily, A. J., *Fracture of Structural Materials*, John Wiley, 1967
Iron and Steel Institute, *Fracture Toughness*, Publication No. 121, 1968
Pratt, P. L. (ed), *Fracture 1969*, Chapman and Hall, 1969
Iron and Steel Institute, *The Effect of Second Phase Particles on the Mechanical Properties of Steels*, 1971
Iron and Steel Institute, *Production and Application of Clean Steels*, 1972
McMahon, C. J., *Materials Science and Engineering*, **25**, 233, 1976
Knott, J. F., Mechanics and mechanisms of large scale brittle fracture in steels. *Materials Science and Engineering*, **7**, 1, 1971
Knott, J. F., *Fundamentals of Fracture Mechanics*, Butterworths, 1973
Hondros, E. D. and Seah, M. P., Segregation to interfaces. *International Metallurgical Reviews*, Review 222, 261, 1977
Taplin, D. M. R. (ed), *Fracture 1977*, University of Waterloo Press, 1977
Olefjord, I., Temper embrittlement. *International Metallurgical Reviews*, **23**, (4), 149, 1978
Briant, C. L. and Banerji, S. K., Intergranular failure in steels: the role of grain-boundary composition. *International Metallurgical Reviews*, **23**, (4), 164, 1978
Residuals, Additives and Materials Properties, Proceedings of a Joint Conference by the National Physical Laboratory, The Metals Society and the Royal Society, London, The Royal Society, 1980

11
Austenitic steels

11.1 Introduction

Some elements extend the γ-loop in the iron-carbon equilibrium diagram (see Chapter 4), e.g. nickel and manganese. When sufficient alloying element is added, it is possible to preserve the face-centred cubic austenite at room temperature, either in a stable or metastable condition. Chromium added alone to a plain carbon steel tends to close the γ-loop and favour the formation of ferrite. However, when chromium is added to a steel containing nickel it retards the kinetics of the $\gamma \rightarrow \alpha$ transformation, thus making it easier to retain austenite at room temperature. The presence of chromium greatly improves the corrosion resistance of the steel by forming a very thin stable oxide film on the surface, so that chromium-nickel stainless steels are now the most widely-used materials in a wide range of corrosive environments both at room and elevated temperatures. Added to this, austenitic steels are readily fabricated and do not undergo a ductile/brittle transition which causes so many problems in ferritic steels. This has ensured that they have become a most important group of construction steels, often in very demanding environments.

11.2 The iron-chromium-nickel system

The binary iron-chromium equilibrium diagram (Fig. 11.1) shows that chromium restricts the occurrence of the γ-loop to the extent that above 13% Cr the binary alloys are ferritic over the whole temperature range, while there is a narrow $(\alpha + \gamma)$ range between 12% and 13% Cr. The ferrite is normally referred to as delta ferrite, because in these steels the phase can have a continuous existence from the melting point to room temperature. The addition of carbon to the binary alloy extends the γ-loop to higher chromium contents (Fig. 11.2), and also widens the $(\alpha + \gamma)$ phase field upto 0.3% C. When carbon is progressively added to an 18% Cr steel, in the range up to about 0.04% C, the steel is fully ferritic (Fig. 11.2a) and cannot be transformed. Between 0.08 and 0.22% C, partial transformation is possible leading to $(\alpha + \gamma)$ structures, while above 0.40% C the steel can be made fully austenitic (Fig. 11.2b) if cooled rapidly from the γ-loop region. The second effect of carbon is to introduce carbides to the structure as indicated in Fig. 11.2:

$$K_0 = M_3C \quad K_1 = M_{23}C_6 \quad K_2 = M_7C_3$$

212 Steels—Microstructure and Properties

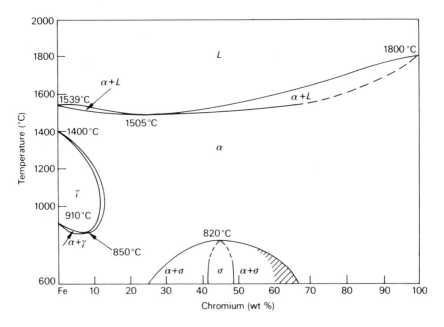

Fig. 11.1 The Fe-Cr equilibrium diagram

In austenitic steels, $M_{23}C_6$ is the most significant carbide formed and it can have a substantial influence on corrosion resistance.

If nickel is added to a low carbon iron–18% Cr alloy, the γ-phase field is expanded until at about 8% Ni the γ-phase persists to room temperature (Fig. 11.3) leading to the familiar group of austenitic steels based on 18% Cr 8% Ni. This particular composition arises because a minimum nickel content is required to retain γ at room temperature. With both lower and higher Cr contents more nickel is needed. For example, with more corrosion resistant, higher Cr steels, e.g. 25% Cr, about 15% nickel is needed to retain the austenite at room temperature. Lack of complete retention is indicated by the formation of martensite. A stable austenite can be defined as one in which the M_s is lower than room temperature. The 18Cr8Ni steel, in fact, has an M_s just below room temperature and, on cooling, e.g. in liquid air, it will transform very substantially to martensite.

Fig. 11.3 also shows that the carbide phase $M_{23}C_6$ exists below about 900°C. However, it goes into solution when the steel is heated to 1100–1150°C and on quenching a precipitate-free austenite is obtained. However, on reheating in the range 550–750°C, $M_{23}C_6$ is reprecipitated preferentially at the grain boundaries.

Manganese expands the γ-loop and can, therefore, be used instead of nickel. However, it is not as strong a γ-former but about half as effective, so higher concentrations are required. In the absence of chromium, around 12% Mn is required to stabilize even higher carbon (1–1.2%) austenite, achieved in Hadfields steel which approximates to this composition. Typically Cr-Mn steels require 12–15% Cr and 12–15% Mn to remain austenitic at room temperature if the carbon content is low.

Like carbon, nitrogen is a very strong austenite former. Both elements, being interstitial solutes in austenite, are the most effective solid solution strengtheners available. Nitrogen is more useful in this respect as it has less tendency to cause intergranular corrosion. Concentrations of nitrogen up to 0.25% are used, which can nearly double the proof stress of a Cr-Ni austenitic steel.

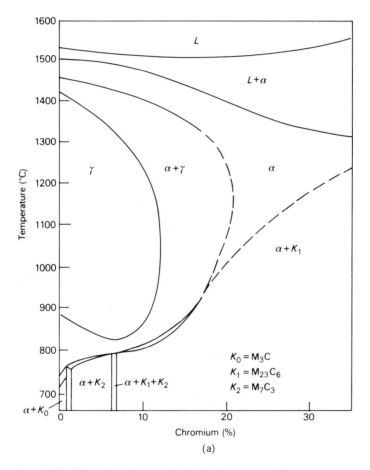

Fig. 11.2 Effect of carbon on the Fe-Cr diagram: a, 0.05C (Colombier and Hoffman, *Stainless and Heat Resisting Steels*, Edward Arnold, 1967)

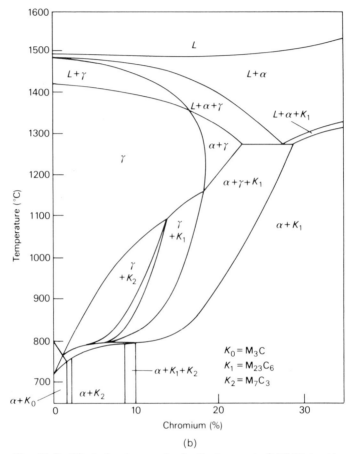

Fig. 11.2 Effect of carbon on the Fe-Cr diagram: b, 0.4C (Colombier and Hoffmann, *Stainless and Heat Resisting Steels*, Edward Arnold, 1967)

One of the most convenient ways of representing the effect of various elements on the basic structure of chromium-nickel stainless steels is the Schaeffler diagram, often used in welding. It plots the compositional limits at room temperature of austenite, ferrite and martensite, in terms of nickel and chromium equivalents (Fig. 11.4). At its simplest level, the diagram shows the regions of existence of the three phases for iron-chromium-nickel alloys. However, the diagram becomes of much wider application when the equivalents of chromium and of nickel are used for the other alloying elements. The chromium equivalent has been empirically determined using the most common ferrite-forming elements:

Cr equivalent = (Cr) + 2(Si) + 1.5(Mo) + 5(V) + 5.5(Al)
              + 1.75(Nb) + 1.5(Ti) + 0.75(W)

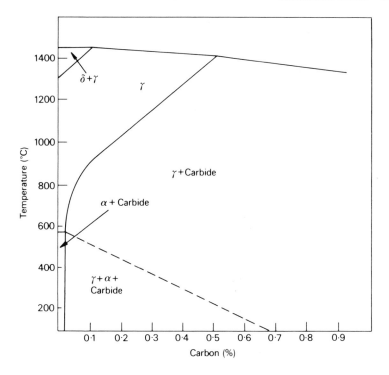

Fig. 11.3 Effect of carbon on the phase diagram for an 18Cr-8Ni steel (Colombier and Hoffman, *Stainless and Heat Resisting Steels*, Edward Arnold, 1967)

while the nickel equivalent has likewise been determined with the familiar austenite-forming elements:

Ni equivalent = $(Ni) + (Co) + 0.5(Mn) + 0.3(Cu) + 25(N) + 30(C)$

all concentrations being expressed in weight percentages.

The large influence of C and N relative to that of the metallic elements should be particularly noted. The diagram is very useful in determining whether a particular steel is likely to be fully austenitic at room temperature. This is relevant to bulk steels, particularly to weld metal where it is frequently important to predict the structure in order to avoid weld defects and excessive localized corrosive attack.

11.3 Chromium carbide in Cr-Ni austenitic steels

Simple austenitic steels usually contain between 18 and 30% chromium, 8 to 20% nickel and between 0.03 and 0.1% carbon. The solubility limit of carbon is about 0.05% at 800°C, rising to 0.5% at 1100°C. Therefore,

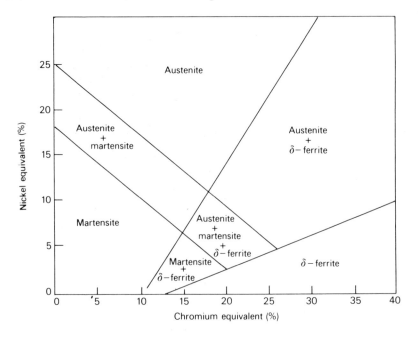

Fig. 11.4 Schaeffler diagram. Effect of alloying elements on the basic structure of Cr-Ni stainless steels (Schneider and Climax Molybdenum Co., *Foundry Trade J.*, 1960, **108**, 562)

solution treatment between 1050 and 1150°C will take all of the carbon into solution and rapid cooling from this temperature range will give a supersaturated austenite solid solution at room temperature. However, slow cooling or reheating within the range 550–800°C will lead to the rejection of carbon from solution, usually as the chromium-rich carbide, $Cr_{23}C_6$, even when the carbon content of the steel is very low ($< 0.05\%$).

This carbide nucleates preferentially at the austenitic grain boundaries as faceted particles (Fig. 11.5) or often as complex dendritic arrays. While such precipitation can have an adverse effect on mechanical properties, in particular low temperature ductility, the most significant result is the depletion of the regions adjacent to the grain boundaries with respect to chromium. This has been revealed directly by microprobe analysis. The surface film in these regions is thus depleted in chromium and as a result the steel is more prone to corrosive attack. Consequently, a classic form of intergranular corrosion is experienced which, in severe cases, can lead to disintegration of the steel. This type of corrosion is also experienced in martensitic chromium steels, e.g. 12% Cr steel, in which grain boundary precipitation of $Cr_{23}C_6$ occurs as well.

$Cr_{23}C_6$ also precipitates within the austenite grains, particularly at higher supersaturations, on dislocations and on solute atom/vacancy clusters.

Austenitic steels 217

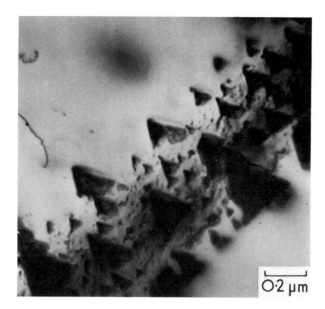

Fig. 11.5 Grain boundary precipitation of $Cr_{23}C_6$ in a 25Cr-24Ni-0.27Ti-0.034C steel aged 5h at 750°C (Singhal and Martin). Thin-foil EM

Both the matrix and the carbide have cubic symmetry, and electron diffraction evidence from thin foil specimens invariably gives the orientation relationship

$$\{100\}_{M_{23}C_6} // \{100\}_\gamma; \langle 100 \rangle_{M_{23}C_6} // \langle 100 \rangle_\gamma$$

The lattice parameter of $M_{23}C_6$ is approximately three times that of austenite, so the electron diffraction patterns are readily identified. The particles usually develop a polyhedral habit, but occasionally in steels deformed at elevated temperatures a more regular cubic morphology is displayed (Fig. 11.6). As the critical temperature range for chromium carbide nucleation and growth is between 500 and 850°C, any process which allows the steel to pass slowly through this temperature range will render it sensitive to intergranular corrosion in service. Welding, in particular, provides these conditions in the heat affected zone (HAZ) leading to localized attack in certain chemical media. It is, therefore, important to have information about the reaction kinetics for the formation of $Cr_{23}C_6$. Being a typical nucleation and growth process, the time-temperature-transformation curve is typically C-shaped with the nose at about 750°C (Fig. 11.7). For some steel compositions the minimum time for the formation of $Cr_{23}C_6$ sufficient to give subsequent intergranular corrosion, i.e. time to achieve *sensitization*, is as short as 100 seconds. Several ways of

218 *Steels—Microstructure and Properties*

Fig. 11.6 Precipitation of $M_{23}C_6$ on dislocations in a 18Cr-12Ni steel after 80h at 700°C under stress (Sully). Thin-foil EM

reducing or eliminating the formation of $Cr_{23}C_6$ are available. The term *stabilisation* is used to describe these processes.

(1) Resolution treatment–after welding, the steel can be reheated to 950–1100°C to allow $Cr_{23}C_6$ to redissolve, and further precipitation is then prevented by rapid cooling to avoid the C-curve.

(2) Reduction of the carbon content–this can be reduced below 0.03% by modern steel-making methods involving oxygen lancing. For complete immunity from intergranular corrosion in 18/8 steels, a carbon level of 0.02% should not be exceeded.

(3) Control of $M_{23}C_6$ reaction kinetics–addition of molybdenum to Cr/Ni stainless steels markedly lengthens the sensitization time. An increase in nickel content has an adverse effect, while increasing chromium has a beneficial effect.

(4) Use of strong carbide-forming elements, Nb, Ti–niobium and titanium form carbides which are much more stable than $Cr_{23}C_6$, so they preferentially combine with the available carbon and thus lessen the opportunity for $Cr_{23}C_6$ to nucleate.

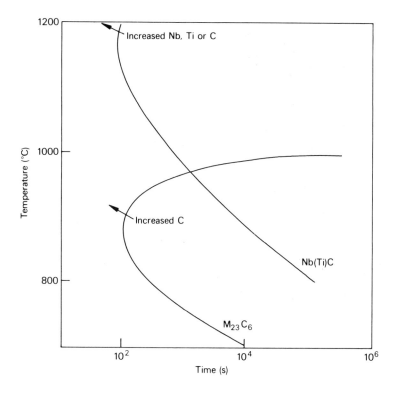

Fig. 11.7 Temperature-time growth curves for $M_{23}C_6$ and Nb(Ti)C in Cr-Ni austenitic steels

11.4 Precipitation of niobium and titanium carbides

In normal practice, sufficient niobium or titanium is added to the steel to combine with all the carbon, the stoichiometric ratios being:

	Ti:C	Nb:C
Atomic weights	48 12	93 12
Ratios	4:1	8:1

However, the additions are in excess of these proportions to allow for some solid solution of Ti or Nb, and for combination with any nitrogen which may be present. Titanium and niobium carbides are much less soluble in austenite than is chromium carbide, so they will form at much higher temperatures as relatively stable particles. These should remain relatively inert during commercial heat treatments involving solution temperatures no higher than 1050°C, and thus minimize the possible nucleation of $Cr_{23}C_6$. However, TiC and NbC have some solubility in austenite at 1050°C and can subsequently precipitate at lower temperatures. During high temperature processes such as welding, these carbides dissolve to a greater extent in austenite and can then reprecipitate at lower temperatures. Therefore, NbC and TiC will not always form inert dispersions, and are often likely to be redistributed by heat treatment. They do, however, have the great advantage of not depleting the matrix of chromium, particularly at sensitive areas such as grain boundaries. The ability to form dispersions of NbC and TiC has a further advantage in that these dispersions can remain very fine at temperatures in the range 500–750°C, and so provide a means of dispersion strengthening austenitic steels to achieve greater strength in this temperature range. The development of creep resistant austenitic steels owes much to the properties of these carbide dispersions.

The formation of NbC and TiC in austenite is most conveniently studied by subjecting the steel to high temperature solution treatment (1100–1300°C), followed by rapid cooling to room temperature. On subsequent ageing in the range 650–850°C precipitation takes place. The carbides are both fcc of the NaCl crystal type with lattice parameters within 2–3% of each other, but differing from that of austenite by 20–25%. They both exist over a range of stoichiometry $MC_{0.6}$–$MC_{1.0}$. Precipitation in each case occurs in several different ways.

Grain boundary Grain boundaries are preferred sites, but because chromium diffuses more rapidly in austenite than does Nb or Ti, $Cr_{23}C_6$ usually forms first (Fig. 11.8a). This emphasizes that NbC or TiC should not be taken into solution if full stabilization is to be achieved. In Fig. 11.7, *TTT* curves for $Cr_{23}C_6$ and (NbTi)C illustrate that, at lower temperatures and shorter times, the chromium carbide forms first, but at longer times it can redissolve and be replaced by (NbTi)C.

Dislocations NbC and TiC nucleate extensively on dislocations (Fig. 11.8b), an important mechanism relevant to the precipitation of equilibrium phases which have not been preceded by GP zone formation. It should also be noted that a significant part of the creep resistance of this group of alloys arises from nucleation of alloy carbides on dislocations generated by deformation at elevated temperatures, e.g. Fig. 11.6. The carbides always have a cube-cube Widmanstätten orientation relationship with the matrix, as do other MC carbides such as VC, TaC. Since the lattice parameter of austenite is 20–25% less than that of the carbides, a flux of vacancies into the precipitates is needed to reduce internal stresses resulting from growth of the particles. Only a few of these vacancies can be quenched in, so carbide particles will grow most readily in situations where further vacancies are generated, e.g. at dislocations or boundaries.

Austenitic steels 221

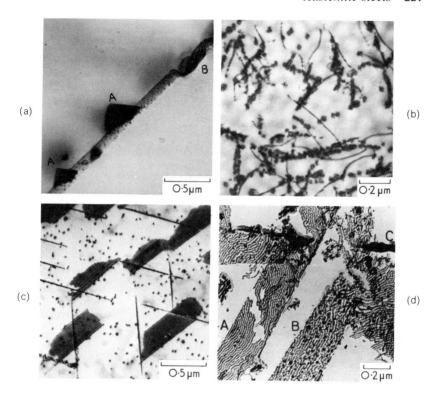

Fig. 11.8 Different modes of precipitation in austenitic steels solution heated between 1150 and 1300°C: a, 25Cr-24Ni-0.27Ti-0.03C, aged 3h at 700°C. Grain boundary precipitation of $M_{23}C_6$ (coarse) and TiC (fine) (Singhal and Martin). Thin-foil EM; b, 18Cr-10Ni-1Ti-0.1C, aged 48h at 700°C. TiC on dislocations (Van Aswegan). Thin-foil EM; 18Cr-12Ni-2Ta-0.1C, aged 25h at 700°C. Matrix and stacking fault precipitation of TaC (Froes). Thin-foil EM; d, 18Cr-12Ni-1.25Nb-0.04N, aged 500h at 700°C. NbN in association with stacking faults A and B, M_6N at B and C (Borland). Extraction replica

Precipitation in association with stacking faults Often NbC, TaC and TiC precipitate on $\{111\}_\gamma$ plates as thin discs which exhibit stacking fault contrast in thin foils in the electron microscope (Fig. 11.8c). These discs grow very substantially on ageing, e.g. at 700°C. Analysis has shown that the discs are formed by the climb of partial dislocations (Frank type), which by climbing generate a continuous source of vacancies. The (NbTi)C precipitate particles nucleate on the partial dislocations and make use of the vacancies in growing, a process which is repeated many times as the partial dislocation escapes from the rows of particles it has nucleated. The final result is a pseudo-Widmanstätten array of discs on $\{111\}_\gamma$ planes, which contain very fine dispersions of (NbTi)C. This complex precipitate morphology can occur side-by-side with normal nucleation on undissociated dislocations.

Matrix precipitation Random precipitation of (NbTi)C in the matrix, not on dislocations, is occasionally observed, but it is the rarest morphology encountered. The particles still exhibit the cube-cube orientation relationship with the matrix, and are apparently nucleated on solute atom/vacancy clusters. Consequently, they are only obtained after heat treatments which result in high supersaturations of vacancies in the austenite matrix, i.e. very high solution temperatures and rapid quenching (Fig. 11.8c). However, there is some evidence that certain elements, e.g. phosphorus, encourage this type of precipitation by trapping vacancies, the phosphorus atoms being 20% smaller than the other atoms in the austenite solid solution, and so cause localized strain fields.

The carbide morphologies have been presented in decreasing order of occurrence. The evidence suggests that this order is dictated by increasing degree of supersaturation, which is a function of the solution temperature. In practice, high solution temperatures can usually be avoided, except in welding, so grain boundary precipitation and dislocation precipitation are the dominant mechanisms observed.

11.5 Nitrides in austenitic steels

In simple austenitic steels the role of nitrogen is largely that of a solid solution strengthening element, although it can replace carbon in $Cr_{23}C_6$. While higher nitrogen concentrations can be maintained without deleterious precipitation than is the case with carbon, in steels with 0.2–0.3% N, Cr_2N can precipitate at grain boundaries, and also within the grains. Exposure of austenitic steels to air at temperatures greater than 600°C can lead to very high ($>1\%$) nitrogen concentrations under the oxide layer, with coarse Cr_2N matrix precipitation, as well as discontinuous lamellar precipitation at grain boundaries. Such regions often lead to cracks under creep conditions.

In the presence of Nb or Ti, more stable nitrides of these elements are formed, which are much less soluble in austenite than Cr_2N. TiN and NbN, isomorphous with the corresponding carbides, have been identified, and also M_6N which can eventually replace NbN during ageing (Fig. 11.8d). These phases can precipitate in the range 650–850°C after rapid cooling from high solution temperatures. They may, therefore, occur as a result of welding or in alloys subject to creep conditions at high temperatures. The modes of nucleation of these nitride phases are similar to those of the corresponding carbides, although there are many morphological differences.

11.6 Intermetallic precipitation in austenite

Austenitic steels, as a class, possess relatively modest mechanical properties, which are largely outweighed by their excellent corrosion resistance in many media. However, it is often desirable to develop higher strength alloys, particularly for use at elevated temperatures where deformation by creep needs to be minimized. Carbide dispersions offer one solution, but the volume fraction of precipitate is limited by solubility considerations and there are also problems associated with high temperature ductility and the stability of the dispersions.

The highly alloyed matrices of many austenitic alloys have allowed the development of intermetallic phases as suitable dispersions to achieve high temperature strength. The most important of these phases is the γ' fcc phase $Ni_3(AlTi)$ first found in nickel-base alloys, with an fcc matrix analogous to austenite, containing titanium and aluminium which can replace each other in the precipitate. The γ' precipitate is obtained in stable austenitic steels, e.g. 20Cr25Ni with an(Al + Ti) content of 1–5%, by quenching from a solution temperature of 1100–1250°C, and ageing in the range 700–800°C. The dispersion developed in this way has two important advantages. Firstly, the precipitate particles have the same crystal structure as the matrix with which they have a cube-cube orientation relationship. Moreover, the lattice parameters are similar, so that the interfaces between precipitate and matrix are coherent and, therefore, of low energy. The familiar Lifshitz-Wagner equation (Equation 8.2) shows that the coarsening rate is directly related to the interfacial energy. Secondly, this type of reaction allows a large volume fraction (30–50%) of precipitate particles to be achieved, the particles being strong, but not catastrophically brittle, cf. sigma phase.

The γ' precipitate normally observed in austenite is spherical when the precipitate is very fine (Fig. 11.9a), and indeed there is evidence for the formation of pre-precipitation spherical zones. However, on prolonged ageing at 750°C, the γ' particles gradually adopt a more complex morphology as they lose coherency with the austenitic matrix (Fig. 11.9b). By varying the ratio of Ti to Al in γ' the coarsening characteristics can be substantially modified. Addition of Al to $\gamma'Ni_3Ti$ decreases the lattice parameter from about 3.590 Å to a value very close to that of the matrix, i.e. around 3.582 Å for 25%Ni 15%Cr, resulting in greater stability of the precipitate. However, complete replacement of titanium lowers the γ' parameter to 3.559 Å, which results in an increase in mismatch parameter. This helps to explain why an (Al + Ti) content of 1–1.5% Al and 3–3.5% Ti was found to be optimum for high strength austenitic steels, resistant to coarsening.

The γ' phase is not the equilibrium phase in austenitic steels with Al and Ti. It is replaced eventually by a coarsely dispersed hexagonal phase $\eta(Ni_3Ti)$ in titanium-containing steels. In steels with a high Al/Ti ratio, the equilibrium intermetallic phase is body-centred cubic β NiAl. Both these

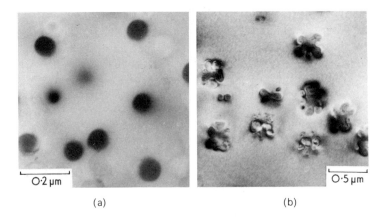

Fig. 11.9 Precipitation of γ' Ni_3Ti in a 21Cr-24Ni-1.3Ti-0.04C steel solution treated at 1150°C: a, 80h at 750°C; b, 800h at 750°C (Singhal and Martin). Thin-foil EMs

phases coarsen excessively, and are undesirable constituents of the microstructure in austenitic creep resistant alloys.

While a number of other intermetallic phases have been observed in austenitic steels, mention will be made only of sigma phase (σ), as it usually has a catastrophic influence on mechanical properties at room temperature. The phase, which is tetragonal in structure, is already present in the binary Fe-Cr system and occurs over a wide composition range between 25 and 60 wt%Cr. In CrNi austenitic steel, σ formation is encouraged when the Cr content exceeds 17%, but is discouraged by increasing the nickel content. The phase forms at austenite grain boundaries and requires, for full development, long-term ageing (up to 1500 hours) at 750°C. However, in some circumstances, σ has been detected in 25 Cr 20 Ni steels after 70 hours at this temperature. The presence of ferrite in the austenite greatly accelerates the formation of sigma, which has been shown to nucleate at the γ/α boundaries (Fig. 11.10). The ferrite, being richer in chromium, tends to be preferentially absorbed during the growth of sigma phase. Elements such as Mo and Ti achieve a further acceleration of sigma formation, e.g. in an 18Cr-8Ni-3Mo-1Ti steel, σ can be formed after only 30 min at 870°C.

11.7 Austenitic steels in practical applications

The commonest austenitic steel is the so-called 18/8 containing around 18%Cr and 8%Ni. It has the lowest nickel content concomitant with a fully austenitic structure. However in some circumstances, e.g. after deformation, or if the carbon content is very low, it may partially transform to martensite

Austenitic steels 225

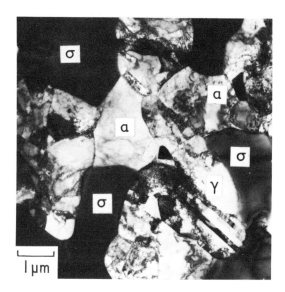

Fig. 11.10 Nucleation of sigma phase at γ/α boundaries (Southwick). Thin-foil EM

at room temperature. Several of the most familiar austenitic steel specifications are given in Table 11.1.

Greater stability towards the formation of martensite is achieved by increasing the nickel content, as illustrated in the 301 to 310 types of steel in the Table 11.1. 18/8 stainless steel owes its wide application to its excellent general resistance to corrosive environments. However, this is substantially improved by increasing the nickel content, and increasing the chromium gives greater resistance to intergranular corrosion. Austenitic steels are prone to *stress corrosion cracking*, particularly in the presence of chloride ions where a few ppm can sometimes prove disastrous. This is a type of

Table 11.1 Some typical austenitic steel specifications

Element	Composition (wt%)						
	AISI type 301	302	304	310	316	321	347
C	0.15 max	0.08 max	0.08 max	0.25 max	0.08 max	0.08 max	0.08 max
N	0.03	0.03	0.03	0.03	0.03	0.03	0.03
Cr	16–18	17–19	18–20	24–26	16–18	17–19	17–19
Ni	6–8	8–10	8–12	19–22	10–14	9–12	9–13
Mo					2–4		
Ti						5 × %C	
Nb							10 × %C
Mn	1.5	1.5	1.5	1.5	1.5	1.5	1.5

failure which occurs in some corrosive environments under small stresses, either deliberately applied or as a result of residual stresses in fabricated material. In austenitic steels it occurs as transgranular cracks which are most easily developed in hot chloride solutions. Stress corrosion cracking is very substantially reduced in high nickel austenitic alloys.

Type 316 steel contains 2–4% molybdenum, which gives a substantial improvement in general corrosion resistance, particularly in resistance to *pitting corrosion*, which can be defined as local penetrations of the corrosion resistant films and which occurs typically in chloride solutions. Recently, some resistant grades with as much as 6.5% Mo have been developed, but the chromium must be increased to 20% and the nickel to 24% to maintain an austenitic structure.

Corrosion along the grain boundaries can be a serious problem, particularly when a high temperature treatment such as welding allows

Table 11.2 Strengthening of austenitic steels at room temperature
(a) Composition

	Composition (wt%)						
	Specification						
Element	304	304(N)	347	347(N)	A286	Unitemp 212	IN744
C	0.08	0.06	0.06	0.08	0.05	0.08	0.05
N	0.03	0.20	0.03	0.20			
Cr	19.0	18.0	18.0	18.0	15.0	13.5	26.0
Ni	10.0	10.0	12.0	11.0	26.0	26.0	6.5
Mo					1.2	1.75	
Ti					2.0	3.0	0.3
Nb			10 × %C	10 × %C			
Al					0.15	0.15	
V					0.30		

(N) = high nitrogen

(b) Mechanical properties

	Specification						
	304	304(N)	347	347(N)	A286	Unitemp 212	IN744
0.2% Proof stress (MNm^{-2})	247	340	247	415	700	920	570
Tensile strength (MNm^{-2})	541	695	556	710	1000	1300	740
Elongation (%)	55	46	50	39	25.0•	23.0■	24

•Aged at 750 C; ■Aged at 700° C
(N) = high nitrogen

precipitation of $Cr_{23}C_6$ in these regions. This type of intergranular corrosion is sometimes referred to as *weld-decay*. To combat this effect some grades of austenitic steel, e.g. 304 and 316, are made with carbon contents of less than 0.03% and designated 304L and 316L. Alternatively, niobium or titanium is added in excess of the stoichiometric amount to combine with carbon, as in types 321 and 347.

The austenitic steels so far referred to are not very strong materials. Typically their 0.2% proof stress is about 250 MNm^{-2} and the tensile strength between 500 and 600 MNm^{-2}, showing that these steels have substantial capacity for work hardening, which makes working more difficult than in the case of mild steel. However, austenitic steels possess very good ductility with elongations of about 50% in tensile tests.

The Cr/Ni austenitic steels are also very resistant to high temperature oxidation because of the protective surface film, but the usual grades have low strengths at elevated temperatures. Those steels stabilized with Ti and Nb, types 321 and 347, can be heat treated to produce a fine dispersion of TiC or NbC which interacts with dislocations generated during creep. One of the most commonly used alloys is 25Cr 20Ni with additions of titanium or niobium which possesses good creep strength at temperatures as high as 700°C.

To achieve the best high temperature creep properties, it is necessary first to raise the room temperature strength to higher levels. This can be done by precipitation hardening heat treatments on steels of suitable composition to allow the precipitation of intermetallic phases, in particular Ni_3(Al Ti). In Table 11.2 the room temperature strength of two alloys in this category (A286 and Unitemp 212) after ageing at 700–750°C is compared with that of the simpler standard austenitic alloys, e.g. 304. It can be seen that the strength is more than doubled by the precipitation reaction.

11.8 Duplex and ferritic stainless steels

In Section 11.2, the importance of controlling the γ-loop in achieving stable austenitic steels was emphasized. Between the austenite and δ-ferrite phase fields there is a restricted $(\alpha + \gamma)$ region which can be used to obtain two-phase or duplex structures in stainless steels (Fig. 11.11). The structures are produced by having the correct balance between α-forming elements (Mo, Ti, Nb, Si, Al) and the γ-forming elements (Ni, Mn, C and N). To achieve a duplex structure, it is normally necessary to increase the chromium content to above 20%. However the exact proportions of α and γ are determined by the heat treatment. It is clear from consideration of the γ-loop section of the equilibrium diagram, that holding in the range 1000–1300°C will cause the ferrite content to vary over wide limits. The usual treatment is carried out between 1050 and 1150°C, when the ferrite content is not very sensitive to the subsequent cooling rate.

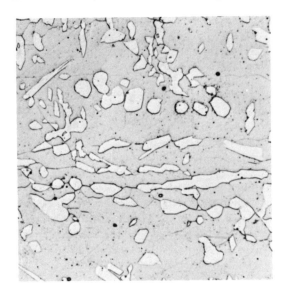

Fig. 11.11 Duplex stainless steel, 26Cr-5Ni-1.5Mo-0.025C, ($\alpha + \gamma$) microstructure (J. Honeycombe). Optical micrograph, ×630

The duplex steels are stronger than the simple austenitic steels, partly as a result of the two-phase structure and also because this also leads normally to a refinement of the grain size. Indeed, by suitable thermomechanical treatment between 900 and 1000°C, it is possible to obtain very fine microduplex structures which can exhibit superplasticity, i.e. very high ductilities at high temperatures, for strain rates less than a critical value. A typical composition, IN744, is shown in Table 11.2 with the mechanical properties at room temperature.

A further advantage is that duplex stainless steels are resistant to *solidification cracking*, particularly that associated with welding. While the presence of δ-ferrite may have an adverse effect on corrosion resistance in some circumstances, it does improve the resistance of the steel to transgranular stress corrosion cracking as the ferrite phase is immune to this type of failure.

There is another important group of stainless steels which are essentially ferritic in structure. They contain between 17 and 30% chromium and, by dispensing with the austenite stabilizing element nickel, possess considerable economic advantage. These steels, particularly at the higher chromium levels, have excellent corrosion resistance in many environments and are completely free from stress corrosion. Typical compositions are shown in Table 11.3.

These steels do have some limitations, particularly those with higher chromium contents, where there can be a marked tendency to embrittlement. This arises partly from the interstitial elements carbon and nitrogen, e.g. a 25% Cr steel will normally be brittle at room temperature if the

Table 11.3 Compositions of some ferritic stainless steels

Element	Composition (wt%)		
	AISI 430	AISI 446	18/2
C	0.06	0.08	0.02
Cr	17.0	25.0	18.0
Mo			2.0

carbon content exceeds 0.03%. An additional factor is that the absence of a phase change makes it more difficult to refine the ferrite grain size, which can become very coarse after high temperature treatment such as welding (Fig. 11.12). This raises still further the ductile/brittle transition temperature, already high as a result of the presence of interstitial elements. Fortunately, modern steel making methods such as argon-oxygen refining can bring the interstitial contents below 0.03%, while electron beam vacuum melting can do better still.

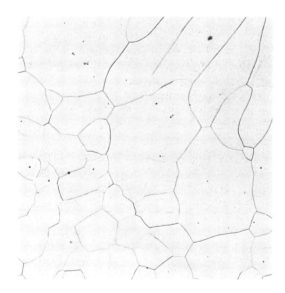

Fig. 11.12 Grain growth in the heat affected zone of a weld in a ferritic stainless steel (J. Honeycombe). Optical micrograph, ×80

The ferritic stainless steels are somewhat stronger than austenitic stainless steels, the yield stresses being in the range 300–400 MNm^{-2}, but they work harden less so the tensile strengths are similar, being between 500 and 600 MNm^{-2}. However, ferritic stainless steels, in general, are not as readily deep drawn as austenitic alloys because of the overall lower ductility. However, they are suitable for other deformation processes such as spinning and cold forging.

Welding causes problems due to excessive grain growth in the heat affected zone but, recently, new low-interstitial alloys containing titanium or niobium have been shown to be readily weldable. The higher chromium ferritic alloys have excellent corrosion resistance, particularly if 1–2% molybdenum is present.

Finally, there are two phenomena which may adversely affect the behaviour of ferritic stainless steels. Firstly, chromium-rich ferrites when heated between 400 and 500° C develop a type of embrittlement (*475° embrittlement*), the origins of which are still in doubt. The most likely cause is the precipitation of a very fine coherent chromium-rich phase (bcc α') arising from the miscibility gap in the Fe-Cr system, probably by a spinodal type of decomposition. This phenomenon becomes more pronounced with increasing chromium content, as does a second phenomenon, the formation of sigma phase. The latter phase occurs more readily in chromium-rich ferrite than in austenite, and can be detected below 600° C. As in austenite, the presence of sigma phase can lead to marked embrittlement.

11.9 The transformation of metastable austenite

Some austenitic steels are often close to transformation, in that the M_s temperature may be just below room temperature. This is certainly true for low-carbon 18Cr8Ni austenitic steel, which can undergo a martensitic transformation by cooling in liquid nitrogen or by less severe refrigeration. The application of plastic deformation at room temperature can also lead to formation of martensite in metastable austenitic steels, a transformation of particular significance when working operations are contemplated. The increase in M_s by cold work is specified by an M_d temperature below which transformation to martensite occurs when the steel is plastically deformed. In general, the higher the alloying element content the lower the M_s and M_d temperatures, and it is possible to obtain an approximate M_s temperature using empirical equations. Useful data concerning the M_d temperature are also available in which an arbitrary amount of deformation has to be specified. Normally this is a true strain of 0.30.

The martensite formed in Cr-Ni austenitic steels either by refrigeration or by plastic deformation is similar to that obtained in related steels possessing an M_s above room temperature. It is usually related to the austenite by the Kurdjumov-Sachs relation.

Manganese can be substituted for nickel in austenitic steels, but the Cr-Mn solid solution then has a much lower stacking fault energy. This means that the fcc solid solution is energetically closer to an alternative close-packed hexagonal structure, and that the dislocations will tend to dissociate to form broader stacking faults than is the case with Cr-Ni austenites. In these circumstances, the martensite which forms first is hexagonal in structure (ε-martensite), with a habit plane $\{0001\}$ parallel to the stacking fault plane $\{111\}_\gamma$. This phase has been shown to nucleate on stacking faults, with the following orientation relationship with austenite:

$\{0001\}_\varepsilon // \{111\}_\gamma$
$\langle 11\bar{2}0 \rangle_\varepsilon // \langle 110 \rangle_\gamma$

This type of martensite forms as parallel-sided plates which can be easily confused with annealing twins, common in fcc matrices with low stacking fault energies. Frequently α' martensite eventually forms, nucleating at the interface between ε and the austenite matrix.

Manganese on its own can stabilize austenite at room temperature provided sufficient carbon is in solid solution. The best example of this type of alloy is the Hadfields manganese steel with 12% Mn, 1.2% carbon which exists in the austenitic condition at room temperature and even after extensive deformation does not form martensite. However, if the carbon content is lowered to 0.8%, then M_d is above room temperature and transformation is possible in the absence of deformation at 77° K. Both ε and α' martensites have been detected in manganese steels. Alloys of the Hadfields type have long been used in wear resistance applications, e.g. grinding balls, railway points, excavating shovels, and it has often been assumed that partial transformation to martensite was responsible for the excellent wear resistance and toughness required. However, it is likely that the very substantial work hardening characteristics of the austenitic matrix are more significant in this case.

In general, fcc metals exhibit higher work hardening rates than bcc metals because of the more stable dislocation interactions possible in the fcc structure. This results in the broad distinction between the higher work hardening of austenitic steels and the lower rate of ferritic steels, particularly well exemplified by a comparison of ferritic stainless steels with austenitic stainless steels. Within the austenitic category, however, there are two factors which influence the extent of work hardening:
(1) the stacking fault energy of the matrix, determined by the composition
(2) the stability of the matrix.

The chromium nickel austenitic steels have stacking faults energies in the range 5–60 mJ m^{-2}, and it would be expected that the highest nickel alloys would show the lowest work hardening as nickel is one of the elements that raises the stacking fault energy of austenite. The elements Cr, Mn, Co, Si, C and N tend to lower the stacking fault energy of austenite. This can be deduced from the greater tendency for annealing twins to occur in austenites rich in these elements. Plastic deformation of such solid solutions not only produces stable dislocation interactions but also, after heavy deformations, many very fine deformation twins, both factors contributing to the high flow stresses observed in deformed alloys. By severe cold working, e.g. up to 80% reduction in wire drawing, the relatively modest yield strengths of ordinary austenitic steels can be raised to over 1200 MNm^{-2}.

However, the largest effect on work hardening rates is undoubtedly the transformation to martensite, as illustrated by the true stress-strain curves of several austenitic steels of decreasing nickel content, i.e. decreasing stability of austenite (Fig. 11.13). By this means yield stresses of well over 1500 MNm^{-2} can be achieved, e.g. a metastable austenite containing 17% Cr, 4% Ni, 3% Mn, 0.1% C after almost complete transformation following 40% deformation at room temperature has a 0.2% proof stress of 1700 MNm^{-2}. It should be noted that the increase in strength is accompanied by a substantial decrease in ductility, so such steels should not be used for deep drawing, an application where stability of the austenite is essential. In contrast, in stretch forming applications unstable austenitic steels can be used because the transformation, by raising the work hardening rate, also increases the extent of uniform straining, as distinct from localized straining, which the steel will undergo prior to failure.

The advantages obtainable from the easily fabricated austenitic steels led naturally to the development of *controlled transformation* stainless steels, where the required high strength level was obtained after fabrication, either by use of refrigeration to take the steel below its M_s temperature, or by low temperature heat treatment to destabilize the austenite. Clearly the $M_s - M_f$ range has to be adjusted by alloying so that the M_s is just below room temperature. The M_f is normally about 120°C lower, so that refrigeration in the range -75 to $-120°$C should result in almost complete

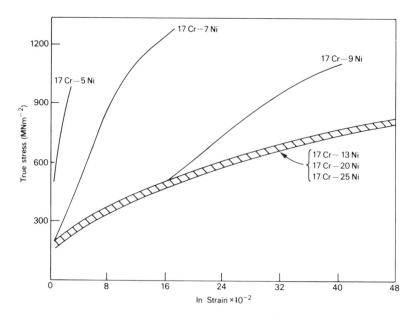

Fig. 11.13 Effect of decreasing nickel content on the stress-strain curves of stainless steels (Pickering, In: *Metallurgical Achievements*, Pergamon Press, 1965)

transformation to martensite. Alternatively, heat treatment of the austenite can be carried out at 700° C to allow precipitation of $M_{23}C_6$ mainly at the grain boundaries. This reduces the carbon content of the matrix and raises the M_s so that, on subsequent cooling to room temperature, the austenite will transform to martensite. This precipitation reaction can be accelerated by designing the steel to include a small volume fraction of δ-ferrite. The δ/γ interfaces then provide very effective nucleating sites for $M_{23}C_6$.

Further heat treatment is then necessary to give improved ductility and a high proof stress; this is achieved by tempering in the range 400–450° C. Typical compositions of these steels and the properties which can be obtained by alternative heat treatments are given in Table 11.4. This category of steels places large demands on metallurgical control, the treatments are complex and the cost high. Consequently, they tend only to be used in critical applications such as highly stressed skins of supersonic aircraft and rocket casings.

Table 11.4 Compositions and properties of controlled transformation steels (after Pickering)

Composition	Heat treatment	0.2% Proof stress (MNm^{-2})	Tensile strength (MNm^{-2})	Elongation (%)
0.1C, 17 Cr, 4Ni, 3Mn	(1) Solution treated 925° C, cold worked 40% reduced, tempered 3h at 450° C	1670	1700	3.5
	(2) Solution treated 950° C, refrigerated at −78° C, tempered 1h at 400° C	1200	1440	19
0.06C, 16.5Cr, 5Ni, 2Mn, 1.5Mo, 2Co, 1Al	(1) Solution treated 1050° C, aged 2h at 700° C, cooled to RT, then aged 4h at 450° C	1270	1430	3
	(2) Solution treated 950° C, refrigerated at −78° C, tempered 4h at 450° C	1240	1520	21
0.07C, 17.5Cr, 3Ni, 2Mn, 2Mo, 2Co, 1Cu	(1) Solution treated 1050° C, aged 2h at 700° C, cooled to RT, then tempered 4h at 450° C	1110	1250	10
	(2) Solution treated 950° C, refrigerated at −78° C, tempered 4h at 450° C	1240	1360	20

RT = room temperature

Another group of steels has been developed to exploit the properties obtained when the martensite reaction occurs during low temperature plastic deformation. These steels, which are called *transformation induced plasticity* (TRIP) steels, exhibit the expected increases in work hardening rate and a marked increase in uniform ductility prior to necking. Essentially the principle is the same as that employed in controlled transformation steels, but plastic deformation is used to form martensite and the approach is broader as far as the thermomechanical treatment is concerned.

In one process, the composition of the steel is balanced to produce an M_d temperature above room temperature. The steel is then heavily deformed ($\sim 80\%$) above the M_d temperature, usually in the range 250–550° C, which results in austenite which remains stable at room temperature. Subsequent tensile testing at room temperature gives high strength levels combined with extensive ductility as a direct result of the martensitic transformation which takes place during the test. For example, a steel containing 0.3% C, 2 Mn, 2 Si, 9 Cr, 8.5 Ni, 4 Mo after 80% reduction at 475° C gives the following properties at room temperature:

0.2% Proof stress	1430 MNm^{-2}
Tensile strength	1500 MNm^{-2}
Elongation	50 %

Higher strength levels (proof stress ~ 2000 MNm^{-2}) with ductilities between 20–25% can be obtained by adding strong carbide forming elements such as vanadium and titanium, and by causing the M_d temperature to be below room temperature. As in the earlier treatment, severe thermomechanical treatments in the range 250–550° C are then used to deform the austenite and dispersion strengthen it with fine alloy carbides. The M_d temperature is, as a result, raised to above room temperature so that, on mechanical testing, transformation to martensite takes place, giving excellent combinations of strength and ductility as well as substantial improvements in fracture toughness.

Like the controlled transformation steels, the TRIP steels require extremely good metallurgical control and are very expensive to make. They are only used in applications where extremely high demands are made on mechanical, as distinct from environmental, properties. They do, however, illustrate how a combination of basic principles can be carefully balanced and controlled to achieve outstanding mechanical properties in alloy steels.

Further reading

Monypenny, J. K. G., *Stainless Iron and Steels*, Volumes 1 and 2, Chapman, 1951

Iron and Steel Institute, *Metallurgical Developments in High Alloy Steels*, Special Report No. 86, 1964.

Colombier, L. and Hochmann, J., *Stainless and Heat Resisting Steels*, Edward Arnold, 1967.

Iron and Steel Institute, *Stainless Steels*, Special Report No. 117, 1969.

Schmidt, W. and Jarleborg, O., *Ferritic Stainless Steels*, Climax Molybdenum, 1974.

Pickering, F. B., Physical metallurgy of stainless steel developments. *International Metallurgical Reviews*, Review 211, 1976

Barr, Robert Q. (ed), *Stainless Steel '77*, Climax Molybdenum Company Conference, London, 1977

Pickering, F. B. (ed), *The Metallurgical Evolution of Stainless Steels*, American Society for Metals/Metals Society, 1979

Index

A_1, A_2, A_3, A_4, temperatures, 28–30
alloy carbides, 60–2
　enthalpies of formation, 60
　fibrous growth, 69, 72
　in tempered martensite, 152–60
　interphase precipitation, 69, 72–3
　nucleation in ferrite, 69, 73–4
　stability, 60
alloying elements,
　α-stabilizing, 55–9
　distribution in steels, 58–62
　effect on equilibrium diagram, 55–8
　effect on kinetics of γ/α transformation, 62–4
　γ-stabilizing, 55–9
　solubility in cementite, 65–6
annealing,
　isothermal, 52
　spheroidize, 52
　subcritical, 53
atmospheres, 15–20
　condensed, 15
Auger spectroscopy of grain boundaries, 205–8
ausforming, 177–82
　applications, 184
　mechanical properties, 184–5
　processing variables, 179–80
　steel composition, 179, 185
　strengthening mechanisms, 181–2
　structural changes, 181
austenite, 30–48
　effect of carbon on lattice parameter, 78–80
　γ-cementite transformation, 33–4
　γ-pearlite transformation, 38–48
　hardness, 98–9
　twin boundaries, 34
austenite-ferrite interfaces,
　curved, 31–3
　planar, 31–3, 37

austenite-ferrite transformation, 31–3
　effect of alloying elements, 62–6
　para-equilibrium, 63
　partition of alloying elements, 62–6
austenite-pearlite interface, 40–1
austenitic steels, 211–26
　chromium carbide in, 215–8
　corrosion resistance, 225–7
　Fe-Cr-Ni system, 211–5
　intermetallic precipitation in, 223–4
　practical applications, 224–7
　mechanical properties, 227
　niobium and titanium carbides in, 219–22
　nitrides in, 222
　specifications, 225
　stacking fault energies, 230–1
　tantalum carbide in, 220–1
　work hardening rate, 232
auto-tempering, 98, 104, 140

Bain strain, 82–4
bainite, 106–19
　carbon concentration gradient near interface, 115–6
　effect of carbon on bainite transition, 114–5
　lower bainite, 108–10
　reaction kinetics, 111–3
　retained austenite, 108
　role of alloying elements, 114–6
　upper bainite, 106–8
bainite start temperature, 112
　effect of carbon, 114–5
　effect of other alloying elements, 114–8
bainitic steels, 117–9
　role of boron, 117–8
　role of molybdenum, 117–8
blue brittleness, 19–20
borides, enthalpies of formation, 60

238 Index

brittle fracture, 186–96, 202–9
  cleavage, 186–96
  intergranular, 202–9
  practical aspects, 194–6
burning, 208–9

CCT (continuous cooling transformation) diagrams, 121–4
  relation to Jominy test, 132–4
C-curve, see TTT curve
carbon,
  atmospheres, 15–20
  effect on bainite formation, 114–5
    on hardenability, 131, 134–5
    on impact transition temperature, 167
    on martensite crystallography, 87–90
    on martensite strength, 98–102
    on phase diagram for 18Cr-18Ni steel, 215
    on tempering, 146–7, 161–3
  solubility in α- and γ-iron, 4–5
  strengthening of iron, 15–20
carbon equivalent, 196
carbon steels, 50–3
  applications, 53
  mechanical properties, 52
  tempering behaviour, 140–8
carbide forming elements, 58–62
carbide pinning of boundaries, 150–2
  in controlled rolling, 168–70
carburizing, 10
cast brittleness, 209
cementite, 8–9, 28–30
  austenite-cementite transformation, 33–4
  cementite-austenite orientation relation, 34
  in lower bainite, 110–1
  in upper bainite, 107–8
  orientation relation with ferrite, 143
  precipitation in ferrite, 7–9
  precipitation in martensite, 143–5, 150–2
Charpy test, 187
chromium carbides,
  during tempering, 157–9
  formed during isothermal transformation, 71–3
  grain boundary precipitation, 216
  in Cr-Ni austenitic steels, 215–8
  pseudo-equilibrium diagram, 61
  sequence in tempering, 157
  TTT curves, 217–19
chromium equivalent, 214
chromium nitride, in austenite, 222
chromium steels,
  ausformed, 184–5
  isothermal transformation, 72–3
  properties of 12% Cr steels, 163–4
  tempering behaviour, 157–9
cleavage fracture, 186–91
  criterion, 191–4
  dislocation mechanisms, 188–9
  effect of fabrication, 196
    grain size, 190–1
    hydrogen, 196
  factors influencing onset, 188–91
  influence of lath packet width, 193
  nucleation by carbides, 189, 193
    by inclusions, 189
    by twins, 189
  practical aspects, 194–6
compressive stresses in surfaces, 135–7
computer determination of ternary equilibria, 59
controlled rolling, 166–74
  boundary pinning, 168
  carbide precipitation, 17
  dispersion strengthening, 172–4
  grain size control, 168–72
  properties of steels, 183–4
controlled transformation stainless steels, 232
coring of dendrites, 11
corrosion,
  intergranular in Cr-Ni steels, 216–8
  pitting, 226
  stress, 208, 225–7
crack nuclei, 188–9
cracks, 186–209
  cleavage, 186–7
  ductile, 196–8
  hot short, 209
  hydrogen embrittlement, 196–7
  intergranular, 202–9
  lamellar tearing, 201
  overheating and burning, 208–9
  quenching, 136–8
  rock candy fracture, 202–4
  stress corrosion, 208
  solidification, 228
  temper embrittlement, 204–8

Index

critical diameter (D_o), 125–6
  ideal critical diameter (D_i), 125–6, 128

Deep drawing steels, 26
delayed cracking, 196
delta ferrite, 1, 211
  effect on stress corrosion of stainless steels, 228
  on nucleation of $M_{23}C_6$, 233
diffusion,
  activation energies of, 7
  during γ/α transformation, 36–7
  during pearlite reaction, 47–8
  of C in α- and γ-iron, 6–8
  of N in α- and γ-iron, 6–8
  of substitutional elements in α- and γ-iron, 7
dislocation locking, 15–20
dislocation pile-up, 23–4
  in crack nucleation, 188–91
dislocation precipitation,
  in austenitic steels, 216, 220–2
  ferrite, 8–9, 74
  tempering, 154–6
dislocations,
  in ausforming, 181
  ferrite, 14, 17–20, 33
  martensite, 86
  stainless steels, 220–1
  tempered steels, 154–6
  screw dislocations in iron, 14
dispersion strengthening, 24–5
  Ashby equation, 25
  of austenitic steels, 215–22
  of HSLA steels, 172–4
  Orowan equation, 24
dual phase steels, 176–7
Dubé classification, 31–2
ductile fracture (fibrous), 196–8
  dimples and voids, 197–8
  micro-necks, 198
  role of carbides, 201–2
  role of inclusions, 199–201
ductility, 199–202
  role of carbides, 201–2
  role of inclusions, 199–201
dynamic recrystallization, 168

Embrittlement,
  475°, 230
  hot short, 209
  hydrogen, 196
  intergranular, 202–9
  overheating and burning, 208–9
  temper, 204–8
epsilon martensite (ε), in austenitic steels, 230–1
equilibrium diagrams
  Fe-C, 28–30
  Fe-Cr, 211–2
  Fe-Cr-C, 212–5
  Fe-Cr-Ni, 214–5
  Fe-X, 55–8
equivalents, Cr and Ni in austenitic steels, 214–5
eutectoid reaction, 28–30, 39
  crystallography of pearlite, 43–4
  kinetics, 44–6
  morphology of pearlite, 39–43
  rate controlling process, 47–8

Fatigue,
  effect of ausforming, 184
  effect of strain ageing, 26–7
  role of inclusions, 201
fatigue limit in steels, 27
ferrite,
  austenite-ferrite transformation, 31–3
  growth by step migration, 33–4, 37–8
  growth kinetics, 36–8
  orientation relationship with austenite, 31
  planar interfaces, 33–4
  precipitation of carbon and nitrogen, 7–10
  quench ageing, 9–10
  Widmanstätten, 31–3
ferritic stainless steels, 228–30
fibrous carbides, 69, 71–2
Fick's second diffusion law, 10
fracture, 186–209
  cleavage, 186–96
  ductile, 196–202
  intergranular, 202–9
fracture toughness, 192, 195
friction stress, 22

Gamma loop, 56–7, 211–2
gamma prime phase, 223–4
  stability of Ni_3Ti, 223

grain boundary
  allotriomorphs, 31–4, 36–8
  cracking, 202–9
  precipitation, 144–5, 156, 216
  segregation, 205–8
grain growth
  during controlled rolling, 168
  in heat affected zone, 229
grain size,
  effect on fracture stress, 193
    hardenability, 129–31
    strength of martensite, 102–3
    yield stress, 22–4, 170–1
  refinement, 53
granular bainite, 108
Griffith criterion for fracture, 191
Grossman test, 124–6

H-coefficient, 125–6
Hadfields steel, 213, 231
Hall-Petch relationship, 22, 102, 170
hardenability, 121, 124–35
  effect of composition, 130–2
  effect of grain size, 129–31
  Grossman test, 124–6
  Jominy test, 126–9
  testing, 124–9
hardenability band (Jominy), 127–9
hardness
  of martensite, 98–9
  of tempered C steels, 146, 148
  of tempered alloy steels, 153–4, 161–4
Holloman-Jaffe parameter, 153
homogenizing, 6, 11
Hultgren extrapolation, 47–8
hydrogen embrittlement, 196

Idiomorphs, 31
impact transition curves, 50–1, 167, 187–8
impurity drag, 63
inclusions, 199–201
  categories, 199
  role in ductile fracture, 197–9
  types of MnS, 200
in situ nucleation
  of alloy carbides, 154–6
  of cementite, 143
intergranular corrosion, in austenitic steels, 216, 226

intergranular fracture, 202–9
  effect of P, Sb, Sn, 205–8
  hot shortness, 209
  overheating and burning, 208–9
  temper embrittlement, 202–8
intermetallic precipitation
  in austenite, 223–4
  γ' Ni_3 (AlTi), 223–4
  sigma phase, 224, 230
internal friction, 6, 103–4
  for determination of soluble carbon and nitrogen, 6
  determining diffusivities, 6–7
  of martensite, 103–4
interphase precipitation, 69, 72–3
  in micro-alloyed steels, 173
  precipitate sheet spacing, 72–3
interstices
  in α-iron, 2–4
  in γ-iron, 2–4
interstitial
  atmospheres, 15–9
  atoms, 4–5
  solid solutions, 4–5
    strengthening, 15–20
    yield point phenomena, 15–20
invariant plane strain, 83
iron,
  bcc structure, 2
  deformation, 12–4
  fcc structure, 2
  flow stress temperature dependence, 13
  slip systems in α-iron, 13
  strengthening, 12 et $seq.$
  work hardening, 12–14
  ε-iron carbide, 8–9, 140–2
  orientation relationship with ferrite (Jack), 141
iron-carbon alloys, 1–20, 28–50
  austenite-cementite reaction, 33–4
  austenite-ferrite reaction, 31–3
  equilibrium diagram, 28–30
  eutectoid reaction, 30, 38–43
  kinetics of transformation, 34–8
iron-chromium system, 211–2
iron-chromium-nickel system, 211–5
isoforming, 178, 182
isothermal transformation
  of alloy steels, 74–5
  of plain carbon steels, 34–6
Izod test, 187

Johnson Mehl equation, 44–5
Jominy test, 126–9
  Jominy hardness-distance curves, 127–9

Lattice invariant deformation, 84
Liberty ships, brittle fracture, 194
Lifshitz-Wagner theory of coarsening, 152–3, 223
local fracture stress (σ_f), 192–4
lower bainite, 108–10
  cementite in, 110
  change of habit plane with temperature, 109–10
  growth of plates, 113
  lath size, 108
  morphology and crystallography, 108–10
Luders bands, 15–8
  extension, 15–6
  propagation, 15–6

M_d temperature, 97, 230–4
manganese sulphide
  formed in overheating and burning, 208–9
  inclusions, Types I, II and III, 199–201
maraging, 164–5
martempering, 138–9
martensite, 20, 76 et seq.
  age-hardening, 98–9
  burst phenomenon, 86, 89, 94
  crystal structure, 78–81
  dislocation density, 86, 102
  habit plane, 77–8, 87–90
  hardness, 98–9
  high carbon, 90
  isothermal, 93–4
  loss of tetragonality, 140–2
  low carbon, 87–8
  medium carbon, 88–90
  pinning of dislocations in, 103–4
  temperature dependence of flow stress, 104
  tetragonality, 78–81, 140–2, 150–1
  twins in, 86–7
  yield strength, 98–102
martensite nuclei, 90–4
  classical theory, 90–3
  Olsen and Cohen theory, 93
martensite start temperature (M_s), 76, 90, 94–5, 232–4

martensite strength, 98–105
  effect of austenite grain size, 102
  effect of carbide precipitation, 98–100
  effect of carbon content, 98–101
  effect of plate size, 102–3
  Fleischer theory, 101
martensitic transformation
  Bain strain, 81–4
  crystallography, 81–6
  effect of alloying elements, 94–6
  effect of deformation, 96–8
  kinetics, 90–8
  mechanical stabilization, 97
  morphology, 86–90
  role of slip and twinning, 84–5, 86–7
  stabilization, 98
  surface displacements, 78–9, 82
  two-shear theory, 84–5
mechanical properties of steels,
  alloy, 147–9
  ausformed, 179–80, 185
  austenitic, 226–7, 232
  carbon, 147–9
  controlled transformation, 232–3
  ferrite-pearlite, 51–3
  high strength low alloy, 170–1, 183–5
  iron-alloy crystals, 21–2
  martensite, 98–103
  mild steel, 23–4
  quench-aged iron, 9–10
  TRIP, 233–4
Metastable austenite, 230–3
  transformation of, 232–3
micro-alloyed steels (high strength low alloy steels), 167–76, 195
  applications, 183–4
  contributions to strength, 174–6
  effect of austenitizing temperature, 170–2
  final grain size, 169–72
  Hall-Petch relation, 170–1
  mechanical properties, 174–6, 183–4
  role of stoichiometry, 173–4
  suppression of yield point, 176
molybdenum carbide
  formed in isothermal transformation, 69–70
  formed in tempering, 155–6, 159–60
  orientation relationship with ferrite, 159

molybdenum carbide (*contd.*)
  transformation to other carbides, 156, 159–60

Nickel equivalent, 215
niobium carbide
  in austenitic steels, 219–22, 226–7
  in micro-alloyed steels, 53, 170–3, 183–4
  in tempered steels, 160–1
nitride formers, 26, 60–1
nitrides
  enthalpies of formation, 60
  in α-iron, 9–10
  in austenitic steels, 222
  in micro-alloyed steels, 168
  solubilities in austenite, 68
nitriding, 10–11
nitrogen
  alloying element, 55, 61, 213, 215
  solubility in α- and γ-iron, 4–5
normalising, 51
nucleation
  in situ, 143, 154–6
  of alloy carbides, 71–3, 154–6
  of iron carbides, 141–3
  of pearlite, 40–1
  on dislocations, 8–9, 18–9, 154–6, 216–21

Ostwald ripening of cementite, 144
overheating, 208–9
oxidation resistance
  Cr-Ni austenitic steels, 226

Partition of alloying elements, 62–6
pearlite,
  alloy carbides in, 66–7
  Bagaryatski relation, 43–4
  crystallography, 43–4
  directional growth, 43
  effect of alloying elements on kinetics, 63–5
  effect on ductility, 50
  effect on toughness, 50
  kinetics, 44–8
  lamellar spacing, 41–3
  morphology, 39–43
  Pitsch-Petch relation, 43–4
  rate controlling process, 47–8
  rate of growth, 44–6, 47–8
  rate of nucleation, 44–6
  strength, 25–6, 49–50
pearlitic steels, 50–3
Peierls-Nabarro stress, 14, 105, 188
phase transformation γ/α, 1–4
pitting corrosion, 226
precipitation,
  alloy carbides during γ/α transformation, 70–3
  during tempering, 152–60
  in austenite, 215–22
  in ferrite, 73–4
  alloy nitrides in austenite, 222
  cementite in martensite, 140–2
  intermetallics in austenite, 223–4
  in ferrite, 165
  ε-iron carbide in martensite, 140–2
  iron carbide in α-iron, 7–9
  iron nitrides in α-iron, 9–10
precipitation on dislocations
  in austenite, 181, 217, 220–2
  in ferrite, 8–9, 69, 73–4, 173
pseudo-binary diagrams, 61–2
pure iron, 1

Quench ageing, 9–10
quench cracking, 136–8
quenching media, 125–6, 137
quenching stresses, 136–9
  reduction of, 138–9

Recovery, in ferrite, 146
recrystallization,
  of austenite, 168–70
  of ferrite, 145–6
retained austenite, 76, 89–90, 108, 140, 142, 150–1
rock candy fracture, 204, 209

Schaeffler diagram, 214–6
secondary hardening, 152–4
  Cr steels, 157–8
  Mo and W steels, 159–60
  V steels, 156–7
segregation,
  in cast steel, 11
  to grain boundaries, 205–7, 208–9
sensitization, 217–8
severity of quench (*H*-coefficient), 125–6
shallow hardening, 135
shape deformation, 82–4

shelf energy, 201
sigma phase (σ), 224, 230
Snoek peak, 6, 104
soaking pits, 11
solid solution,
   elastic behaviour of solute and solvent, 22
   Hume-Rothery size effect, 22
   interstitial, 4–5
   substitutional, 4–5, 21–2
solid solution strengthening, 15–24
   interstitial, 15–20
   substitutional, 21–2
solidification cracking, 228
solubility products of carbides and nitrides, 68–9, 168–9
stabilization,
   austenitic steels, 217–18
   during martensite transformation, 98
   mechanical, 97
stacking fault energy of austenite, 230–1
stacking faults in austenitic steels, 220–2
stainless steels, 211–30
   austenitic, 211–27
   carbides in, 215–22
   controlled transformation, 232–3
   corrosion behaviour, 216–7, 225–8, 230
   duplex, 227–8
   ferritic, 228–30
   intermetallic precipitation in, 223–4
   nitrides in, 222
steels (industrial),
   ausformed, 179–80, 185
   austenitic, 224–7
   bainitic, 117–9
   carbon, 50–3
   controlled transformation, 232–3
   hardenability, 121–39
   high strength low alloy (HSLA), 183–4
   mechanical properties of
      alloy steels, 161–4
      carbon steels, 147–8
   thermomechanical treatments, 170–1, 183–5
   TRIP, 233–4
steels for low temperatures, 195
stoichiometric ratio for TiC and NbC, 219

stoichiometry, effect on precipitation of TiC, 174
strain ageing, 15–18
   dynamic, 18, 20
strengthening of iron,
   dispersion, 9–10, 24–5
   grain size, 22–4
   solid solution (interstitial), 15–20
   solid solution (substitutional), 21–2
   work hardening, 12–4
stress,
   critical local fracture stress, 192–4
   effective shear stress (Fe crystals), 14
   effective shear stress (crack nucleation), 192
   flow stress of iron, 13–4
   friction, 22
   yield, 15–9
stresses in heat treatment, 137–8
stress corrosion, 208, 225–6, 228
stress intensity factor (K), 187–8, 192
   critical (K_{IC}), 192
superplasticity in duplex stainless steels, 228

TTT (time temperature transformation) diagrams,
   alloy steels, 74–5, 121–4
   plain carbon steel, 34–6
temper embrittlement, 204–8
   inter-element effects, 205–6
   optimum temperature range, 207–8
   role of carbide particles, 206–7
   segregation of alloying elements, 205–7
   use of Auger spectroscopy, 205–6
tempered martensite embrittlement, 150, 161
tempering, 140–65
   alloy steels, 148–65
   chromium steels, 157–9
   molybdenum and tungsten steels, 159–60
   plain carbon steels, 140–7
   vanadium steels, 156–7
tempering of alloy steels, 148–65
   carbide transformation, 157–60
   coarsening of cementite, 150–2
   complex steels, 160
   mechanical properties, 161–4
   nucleation of alloy carbides, 154–6
   retained austenite, 150–1

tempering of alloy steels (contd.)
  role of dislocations, 154–6
  secondary dispersions, 160–1
  temper embrittlement, 204–8
  tempered martensite embrittlement, 150
tempering of carbon steels,
  activation energies, 145
  coarsening of cementite, 144–6
  effect of carbon, 146–7
  grain boundary cementite, 144
  mechanical properties, 147–9
  nucleation and growth of Fe_3C, 143–5
  precipitation of ε-iron carbide, 140–2
  recrystallization of ferrite, 146
temper rolling, 26
tetragonality of martensite, 78–81
  changes during tempering, 140–2, 150–1
thermal stresses, 136–7
thermionic emission microscopy, 36–7
thermomechanical treatment, 166–85
  ausforming, 177–82
  controlled rolling, 166–74
  high temperature, 182
  isoforming, 182
titanium carbide,
  in austenitic steels, 219–22, 226–7
    micro-alloyed steels, 53, 172–4, 183
  TTT curves for precipitation, 219–20
transformation stresses, 2, 136–7
transition temperature (ductile/brittle (T_c)), 51, 167, 187–8
  effect of grain size, 190–1
TRIP steels, 233–4
tungsten carbide,
  formed during tempering, 159–60
  orientation relationship with ferrite, 159
  transformation to other carbides, 159–60
twinning,
  in austenite, 231
  in martensite, 86–7

Upper bainite, 106–8
  cementite in, 108
  growth of plates, 112

lath size, 107
ledge movement, 108
morphology and crystallography, 106–8
underbead cracking, 196

Vacancies,
  in austenitic steels, 220–2
  in coarsening of Fe_3C, 145
  trapping of by phosphorus, 222
vanadium carbide,
  formed during isothermal transformation, 72–4
  formed during tempering, 156–7
  in micro-alloyed steels, 172–6
  orientation relationship with ferrite, 156

Welding,
  carbon equivalent, 196
  effect of carbon, 53, 134, 167
  HSLA steels, 183
  hydrogen embrittlement, 196–7
  lamellar tearing, 202
  Schaeffler diagram, 214
  solidification cracking, 228
  stainless steels,
    ferritic, 230
    austenitic, 217–8, 226
    duplex, 228
  weld zone cracks, 194
weld-decay, 226
Wever classification, 55–6
Widmanstätten precipitation,
  alloy carbides, 72, 156–9, 220–1
  cementite, 32–4, 143
  ferrite, 31–3, 74

Yield drop, 19
yield point, 15–20
  Cottrell-Bilby theory, 15–8, 188
  Gilman-Johnson theory, 19–20, 189
  lower, 15
  Luders band, 15
  in mild steel, 18
  strain ageing, 15
  stretcher strains, 15
  upper, 15

Zener-Hillert equation, 37

Sturmgeschütz 40 (L48)

Horst Scheibert

The *Sturmgeschütz* 40 in its final configuration (*Ausführung* G with *Saukopfblende, Zimmerit* plaster, rails for hanging *Schürzen*, cast deflector in front of the commander's cupola and machine-gun shield). In this configuration it was in action from 1944 to 1945.(BA)

SCHIFFER MILITARY HISTORY
West Chester, PA

Translated from the German by David Johnston.

Copyright © 1991 by Schiffer Publishing Ltd.
Library of Congress Catalog Number: 91-60860.

All rights reserved. No part of this work may be reproduced or used in any forms or by any means—graphic, electronic or mechanical, including photocopying or information storage and retrieval systems—without written permission from the copyright holder.

Printed in the United States of America.
ISBN: 0-88740-310-7

This title was originally published under the title, *Sturmgeschütz 40 - Der beste Panzerjäger*, by Podzun-Pallas-Verlag GmbH, 6360 Friedberg 3 (Dorheim). ISBN: 3-7909-0192-X.

We are interested in hearing from authors with book ideas on related topics.

Published by Schiffer Publishing, Ltd.
1469 Morstein Road
West Chester, Pennsylvania 19380
Please write for a free catalog.
This book may be purchased from the publisher.
Please include $2.00 postage.
Try your bookstore first.

The G Model was also built in this configuration — with the angular mantelet — until the end of the war.(BA)

Sturmgeschütz 40
(Sd.Kfz. 142) Ausführungen F, F/8 and G

The Sturmgeschütz III was originally intended as a support weapon for attacks by non-armored infantry. It proved to be inadequate in late 1941 with the appearance of the Russian T 34, KV I and KV II tanks, with their powerful armor, high maximum speeds and long-ranging trajectory from long-barrelled guns. The short 75mm gun (Stuk L/24) of the StuG. III remained effective against "soft" and less maneuverable targets but had little effect on the new Russian tanks. A longer gun with better characteristics was what was needed, and a version of the Sturmgeschütz III was developed — relatively quickly — with the long-barrelled gun (Stuk 40). Vehicles armed with the new gun received the following designations:
— Sturmgeschütz 40, *Ausführung* F with the Stuk 40/L43
— Sturmgeschütz 40, *Ausführung* F/8 with the longer Stuk 40 L/48
— Sturmgeschütz 40, *Ausführung* G with the Stuk 40 L/48 and
— Sturmgeschütz 40, *Ausführung* G with the Stuk 40 L/48 in *Saukopf* mantlet

The designation "Sturmgeschütz III" remained in use until the end of the war. The number "40" was intended merely to indicate that the vehicle in question was a Sturmgeschütz III with the longer *Sturmkanone* 40. The two G Models were easily distinguished from the F and F/8 by their commander's cupolas. Distinguishing between the F and F/8 Models, especially if the front end of the vehicle is not clearly visible in the photograph, is more difficult. Shortages of replacement parts, the change to the Stuk 40 L/48 and improvisation by the units resulted in various combinations of armament and equipment — as with most other German combat vehicles. The last two types were built until the end of the war. In total, about 8,000 of the F and G Series were built. Together with the A-E Series about 9,000 Sturmgeschütze were built. No other combat vehicle of the Wehrmacht reached this production figure.

It was only special units such as "Grossdeutschland" and several of the Waffen-SS which had organic assault gun units. Usually they were concentrated in independent battalions and brigades which were attached to corps or armies for inclusion in the main effort of an attack or — especially in the final years of the war — to master a crisis by being employed against enemy armored forces which had broken through. With the appearance of greater masses of enemy tanks the Sturmgeschütz became more and more a defensive weapon — a tank destroyer, a task, as the word "*Sturm*" would indicate, for which it had not originally been intended. Because of its greater ease of manufacture and the lesser demands it made on materials, from 1944 assault guns began to take the place of battle tanks in Panzer Divisions and tank destroyers in Infantry Divisions. As a result of this process there were Panzer Divisions which, in addition to having a tank battalion, also had an assault gun battalion, and tank destroyer battalions which were wholly equipped with assault guns.

With this the confusion reached its high point: the chassis and gun of the Sturmgeschütz originated from the Panzer arm, from 1942 it was forced more and more into the role of a tank destroyer, and finally, from 1944 on, it was employed as a "battle tank." However, in spite of all of this, it was still a piece of artillery (Sturmartillerie!) and until the end of the war its production and training were superintended by that branch of the service. As a result of this situation, Inspector General of Armored Troops Guderian, despite his overall responsibility for all of the Army's armored forces, never had any influence over the design or manufacture of assault guns or the training of their crews. It was an unfortunate situation with negative consequences!

Sturmgeschütz 40
Ausführung F with the Stuk 40 L/43

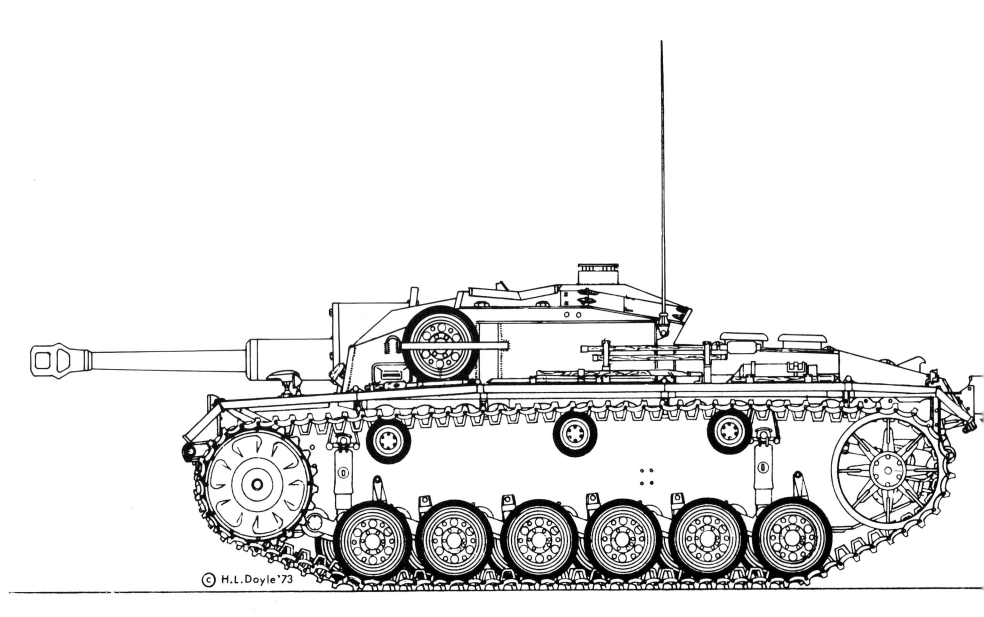

The first version of the Sturmgeschütz III with the longer gun was the *Ausführung* F with the Stuk 40 L/43.

The F Model differed from the F/8 which followed in its somewhat shorter gun, which was usually seen with the globular, single-baffle muzzle brake. Since this gun was also produced with the later double-baffle muzzle brake, it is often difficult to decide whether a Model F or F/8 is illustrated. In cases where doubts exist only a view of the front of the vehicle can help. If the Sturmgeschütz has bolted supplemental armor plates on the nose and driver's plate, it is always an *Ausführung* F/8.

Above and below: This is clearly recognizable as an *Ausführung* F by the presence of the projecting boxes on the sides of the superstructure and the absence of bolted-on supplemental armor. The different light arrangements on these two vehicles is very interesting. The photograph above shows a prototype, that below a vehicle of the *Sturmartillerie-Lehrbataillon*.

Above: Probably an *Ausführung* F; recognizable by the somewhat shorter barrel (L/43) and the globular, single-baffle muzzle brake. Here the Inspector General of Panzer Troops, *Generaloberst* Guderian, is addressing men of the *Reichsarbeitsdienst* (Reich Labor Service).

The two upper photos also show F Models, as the vehicle on the left has the older muzzle brake and that on the right lacks supplemental armor. Noteworthy in the photograph to the left are the vehicle's unusual camouflage scheme and the name "Erika" beside the *Balkankreuz*. Sitting on the assault gun is an umpire (note the white bands on his sleeve and cap). The photograph was taken during an exercise. The gun is fitted with the later double-baffle muzzle brake. Since the front of the vehicle cannot be seen, this may be an F/8 Model.

The assault gun in the above photo carries spare roadwheels on both sides on the forward sections of the track shields. Such a mounting was typical of the F and F/8 Models; on the later *Ausführung* G, with its revised superstructure, this was no longer possible.(1 x BA)

One of the earliest photographs (here a prototype) of the *Ausführung* F. It shows clearly all of the typical features of this version, especially the unreinforced bow and driver's plates.

The carriage of spare roadwheels in front of the left box-shaped protrusion of the superstructure identifies this vehicle as an *Ausführung* F (or F/8).(BA)

This is clearly an *Ausführung* F, as the driver's plate shows no reinforcement — with the exception of a few spare track links.

Southern Russia, August 1942. A good photograph of an *Ausführung* F with the typical spare roadwheel carriage. In addition to the older, covered headlights, this vehicle has a *Notek* light on the left track shield.(BA)

Rear view of one of the first production examples of the *Ausführung* F. Clearly visible is the ventilator, which had become necessary because of the greater accumulations of powder gas from the longer 75mm shells. Later, on the *Ausführung* G, the ventilator was lowered and even later moved to the rear wall of the superstructure.

Sturmgeschütz 40
Ausführung F/8 with the Stuk 40 L/48

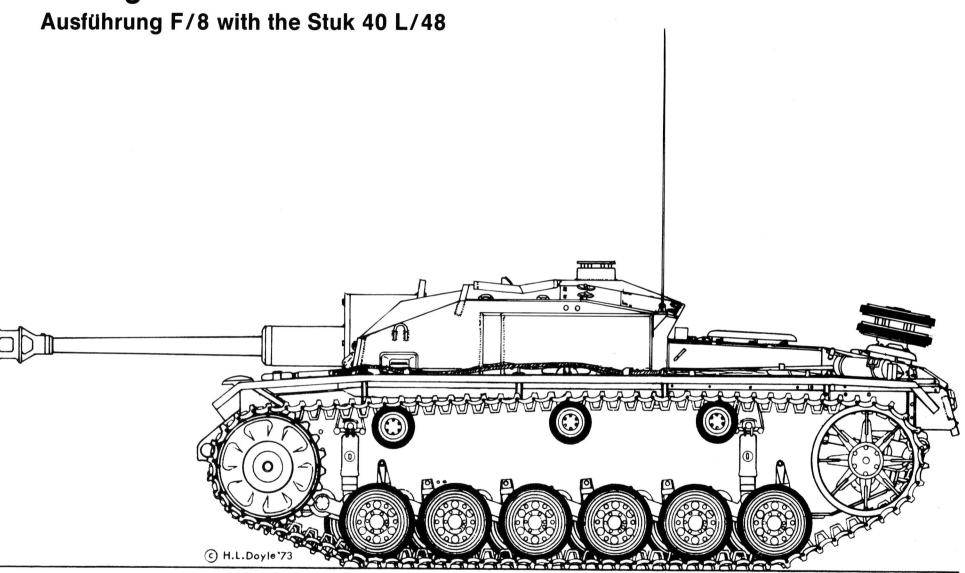

In the side view the F/8 is very similar to the *Ausführung* F — only the gun was longer, and this was usually noticeable only when the two versions stood side by side.

The F/8 differed from the F Model in its somewhat longer gun (often difficult to determine, especially since the odd L/48 was to be found with the earlier globular muzzle brake). If the vehicle has bolted or welded supplemental armor on the bow and driver's plates, it is always an F/8.

The F/8 clearly differed from the following *Ausführung* G, whose initial production lots still had bolted armor, by:
— the absence of a commander's cupola and
— the vertical superstructure side walls with protruding boxes on both sides first seen on the E Model. The G Model had a somewhat wider superstructure with slightly sloping side walls.

Right above: Early examples of the *Ausführung* F/8 had the supplemental 30-mm armor bolted onto the bow and driver's plates.

Right: Later, and on greater numbers of vehicles, these plates were welded on. There were also various combinations to be seen on individual vehicles; usually with welded bow plates and a bolted-on driver's plate. Here, however, both are welded on. With the additional 30-mm armour plate the vehicle's frontal armour was 80 millimeters thick.

Above: The globular muzzle brake indicates that this is an *Ausführung* F. Interesting is the elephant unit insignia next to the *Balkankreuz*. On the Don, near Rostov, 1942.

Above right: A good top view. The star on the side projection indicates that this *Sturmgeschütz* has been captured by U.S. forces.

Right: While only a few F-Models carried *Schürzen*, almost all F/8s carried the metal side panels for protection against hollow-charge shells.

Above left and left: F/8s in winter camouflage. The vehicle on the left has welded supplemental armor, that on the right has bolted-on plates. In addition, the vehicle on the left shows an unusual style of stowing its spare roadwheels, while that on the right has the older type of winter tracks, or *Ostketten* (in which extensions were fitted to the normal track links. Later *Ostketten* had one-piece links).

Left: Of particular interest in this photo are the reinforced (usually with concrete) sloping armor plates on either side of the gun. The tree trunks the vehicles are carrying served as aids in difficult terrain, being placed square beneath the tracks when needed.

A Sturmgeschütz 40 in action in the Ukraine in 1942 while attached to the 19th Panzer Division. This vehicle has welded supplemental armor and the early type of *Ostketten*.

On this vehicle the supplemental armor has been welded over the driver's plate. More often, however, this was bolted on — in contrast to the bow plates — as may be seen on most of the Sturmgeschütze in this volume. (BA)

The presence of *Schürzen* strongly suggests that this is an *Ausführung* F/8. There were variations in the number and sizes of the panels which made up the apron, as a comparison with the above picture will show.

Sturmgeschütz
Ausführung G with Stuk 40 L/48

Typical for the *Ausführung* G are the commander's cupola and the absence of the box-shaped projections on the sides of the superstructure. The latter have been incorporated into the revised, enlarged superstructure of the G-Model.

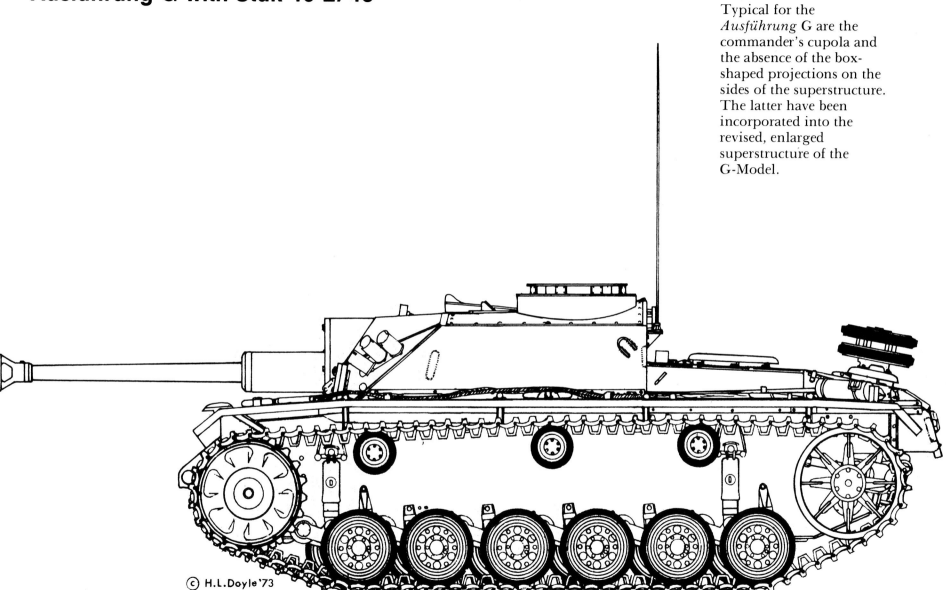

The G Model began reaching the units in 1943 and remained the major production variant until 1945. This version also displayed the bolted — later welded — supplemental armor on the bow and driver's plates introduced on the F/8. Initially it was still equipped with a driver's binocular vision device (recognizable by the two holes above the driver's visor), although this was deleted from later production batches. The roof-mounted ventilator in the middle of the roof plate, introduced on the *Ausführung* F, was made shallower and later moved to the rear wall of the superstructure. Found on most vehicles in front of the loader's hatch was a folding machine-gun shield.

The most obvious feature of the *Ausführung* G, however, was the commander's cupola with its eight periscopes.

Field modifications resulted in most G Model vehicles receiving skirt armor (*Schürzen*) for protection against hollow-charge shells and a coat of *Zimmerit* against magnetic mines. Despite the later introduction of a cast gun mantlet (*Saukopfblende*), for production reasons the *Ausführung* G was also built with the angular mantlet until the end of the war.

Both photographs on this page illustrate the *Ausführung* G. The upper photograph is most interesting as it shows a vehicle with a lower, roof-mounted ventilator (rare!) and the two holes for the driver's binocular vision device. What is more, all of the vehicle's supplemental frontal armor is bolted on.

In contrast, the Sturmgeschütz in the photo on the right has welded supplemental armor (as did the vast majority of G models) and no ventilator on the roof.

A good view of the revised superstructure of the *Ausführung* G with its smooth, slightly sloping sides and the ventilator mounted on the rear wall. The position of the hatch cover on the commander's cupola suggests that this was a revolving cupola (later deleted in favor of a fixed one).

This Sturmgeschütz 40 features the very commonly seen combination of application methods for the supplemental armor: bow plates welded, driver's plate bolted. The armor plate around the driver's vision block is the later type lacking the two ports for the binocular vision device.(BA)

The vehicles in both photographs on this page are fitted with six-section *Schürzen* (as opposed to the more common four-part aprons). In combat it was usual for steel helmets to be worn, as the commander and loader often found themselves with their upper bodies outside the vehicle where they were vulnerable to shrapnel and small arms fire.

The tarpaulins over the gun mantlet, muzzle brake and sighting mechanism protected these sensitive parts in bad weather. (2 x BA)

All three photographs on this page show G-Models.

The vehicles in the two upper photographs wear an improvised winter camouflage finish. The Sturmgeschütz on the left is equipped with discharger cups for smoke candles. These were more commonly found on tanks than assault guns.

While the photo above shows a Sturmgeschütz with welded bow armor, those in the other two have bolted plates. (3 x BA)

The Sturmgeschütz above has concrete reinforcement on both sides of the gun — of questionable value against large-caliber shells.

The vehicle above right is wearing an unusual camouflage scheme.

The photograph on the right was taken in the Baltic Provinces on 14 July 1944. The vehicles have the later-style, wider *Ostketten*, which were actually intended for winter use, and are carrying plenty of spare track as further supplemental armor. The asymmetrical, one-piece *Ostkette* represented a great improvement over the earlier type.(3 x BA)

Above left: January 1945 in Hungary. A Sturmgeschütz 40 (G) of the Waffen-SS with *Zimmerit* coating seen transporting Grenadiers. The machine-gun may be seen protruding through the cut-out in the protective shield.(BA)

Above: An *Ausführung* G with steel return rollers (without rubber tires). These were adopted towards the end of the war due to material shortages. In the left background is a *Bergetiger* (recovery vehicle on Tiger chassis).(BA)

Left: A Sturmgeschütz 40 passes Staraya Russia in 1943.

The lack of protruding boxes on the superstructure sides and the typical hatch cover on the commander's cupola identify this vehicle as an *Ausführung* G. The absence of rails for hanging *Schürzen* suggest that this photograph was taken in 1943 and that the vehicle was from an early production batch of this variant.(BA)

Schürzen rails are discernable in this photograph. The two antennas suggest a command vehicle. In the foreground a surrendering Russian soldier crawls from his foxhole.

16 to 18 victory rings are painted on the gun barrel of this successful Sturmgeschütz.(BA)

With *Schürzen*, spare track links, a tree trunk for difficult terrain and towing cables already attached, this vehicle is ready for anything. To the left of the main gun (as seen by the viewer) are welded steel rods for stowing additional spare track.(BA)

View of the driver's hatch and visor.(BA)

This bird's eye view is very interesting. It shows the inner side of the folded-down machine-gun protective shield, the guide rail of the sighting mechanism and the housing for the barrel recuperator and the recoil cylinder behind the gun mantlet. The individual track links around the cupola were intended to offer additional protection. The widespread use of spare track links as protection was a questionable practice, however, as the links consisted of weaker material with very irregular surfaces and thus acted more as "shot traps." (BA)

Of these three photographs of G Models, the one on the right is noteworthy as it shows a Sturmgeschütz 40 (G) in the Finnish Army Museum, which has been opened to show the interior of the fighting compartment (the Finnish Army was also a user of this version). Also discernable on this vehicle are the deflector (initially made of concrete, later steel) in front of the cupola and steel return rollers. The purpose of the device above the forward shock absorber is unknown.(BA)

Above: In use with a driving school — here during the administering of a driving test.

Above left: The detachable Notek light has been removed and stowed inside.

Left: This Sturmgeschütz 40 wears the later style *Ostketten*. Due to a shortage of regular tracks the vehicle has retained the winter tracks in summer.

Different deflectors are visible in front of or around the commander's cupola on all of the vehicles illustrated on this page. As none of the gun mantlets are visible, these may be *Ausführung* G vehicles with the *Saukopfblende*.

The Sturmgeschütz 40 in the photo above ran over a mine near Weeze (on the Dutch border), was abandoned and is seen here being inspected by English soldiers.(2 x BA)

Sturmgeschütz 40
Ausführung G with Stuk 40 and Saukopfblende

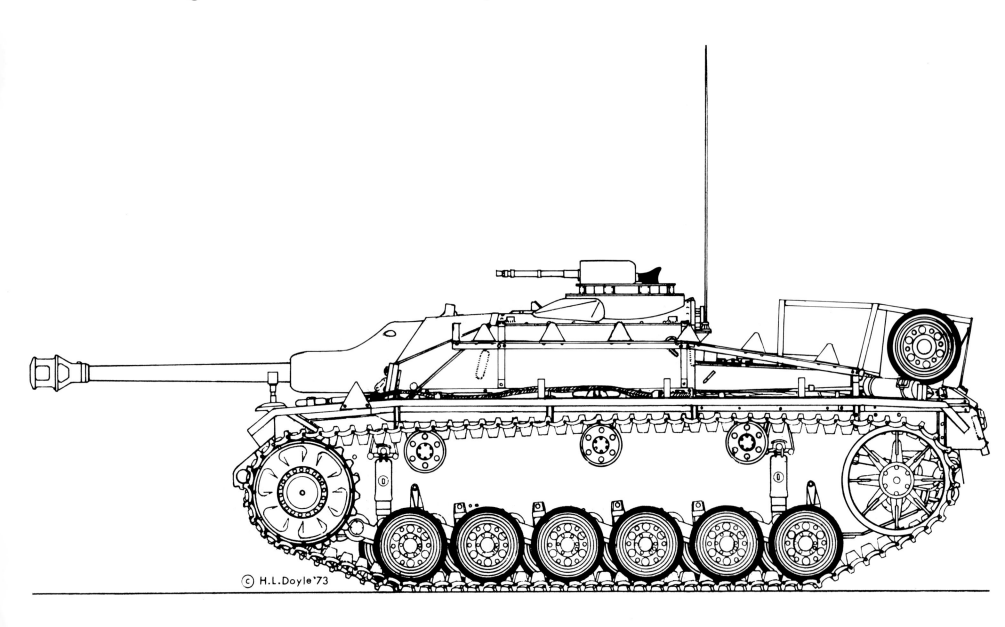

This was the last and most developed version of the proven Sturmgeschütz III. It was delivered from the factory with *Schürzen* and usually with a coating of *Zimmerit*.

With the adoption of the shot-deflecting *Saukopfblende* (cast steel) there were now — following the transfer of the ventilator from the roof to the rear wall of the superstructure and the deletion of the side boxes in favor of smooth, sloping, armored walls — fewer "shot traps" on this version. Vehicles from the final production batches featured a deflector (usually of cast steel) in front of the commander's cupola, a co-axial machine-gun in the *Saukopf* mantlet and a narrower protective shield (originally designed for the "Hetzer") for the loader's machine-gun.

The difference between the *Saukopf* mantlet and the earlier, angular mantlet is obvious. The Sturmgeschütz seen here is a command vehicle (two antennas). Of interest are the two somewhat higher middle *Schürzen* and the deflector in front of the cupola.(BA)

Above left: A Sturmgeschütz 40 (*Saukopf*) with the tactical insignia of the panzer units. Supplemental bow armor was no longer seen on later G Models as the basic armor thickness had been increased to 80mm. However, the supplemental armor was to be seen on the driver's plate until the end of the war — generally bolted on.

Above: The only indication that this is a G Model with the *Saukopf* mantlet is the more conical shape of the armoured gun sleeve.

Left: Driving practice in Holland, 1944.(3 x BA)

It was not at all unexpected that the *Saukopf* mantlet was soon painted to resemble a boar's head.(BA)

Above left: Training on an exercise grounds. In the foreground is a Grenadier in a foxhole with a *Panzerfaust*.

Above: Interesting features of this vehicle are *Schürzen* on only one side, the deflector in front of the commander's cupola and the driver's plate supplemental armor which is not bolted on the left side (as seen by the viewer) as the basic armor thickness there had been increased to 80 mm.

Left: This Sturmgeschütz has steel return rollers, indicating that it could not have been built earlier than the second half of 1944. (3 x BA)

This photograph was probably also taken during training. Interesting are the crosswise patterns of wire strung over the sloping armor plates on both sides of the gun. These provided a better base for the application of camouflage material.(BA)

The two photos on the left show Sturmgeschütze 40 (G) without *Zimmerit* coating — above in the East during the winter of 1944/45, below in the Ukraine in 1944.

The photo above was also taken in southern Russia.

Facing page: This Sturmgeschütz has steel return rollers. In this view it can clearly be seen that the folding sections at the front (and rear) of the mudguards have also received a coating of *Zimmerit*.

Schürzen differed in the number of panels, their size and shape, as well as in the way they were hung (rigidly or, later, loosely). The *Schürzen* on this vehicle are fixed rigidly in place. The position used here for the MG 42 with its bipod on the roof was quite anomalous.(BA)

Above: A Sturmgeschütz in France before the invasion, 1944.

The two photos on the right show vehicles without *Zimmerit* coating; the one in the upper photograph has steel return rollers. (3 x BA)

On later versions the *Zimmerit* coating was usually applied in a waffle-plate pattern. Here the crew are wearing the black Panzer uniform, suggesting that this Sturmgeschütz belongs to a Panzer Division. (BA)

From these photographs it can be seen that the later production lots of the *Ausführung* G (in general those with the *Saukopf* mantlet) no longer received welded supplemental bow armor. The vehicle's hull was now built with 80mm bow armor plates.(3 x BA)

Sturmgeschütze in Germany (above left), near Lake Ilmen (above), and in the Ukraine (left).

Of particular interest in the photograph on the left is the *Balkankreuz* on the skirt armor. This is unusual for two reasons. First, from 1944 the national marking was rarely applied as it was not visible from the long ranges at which combat took place and, second, because there seemed little sense in applying the marking to the *Schürzen*, which were often lost.

The white "K" was part of a continuous lettering sequence of vehicles in this unit. (3 x BA)

45

Which versions are these?(4 x BA)

Technical Data

75mm Sturmgeschütz 40
(Sd.Kfz. 142/1) Ausführung F/8 and G

Engine	Maybach HL 120 TRM	Range	Roads — 155 km	Loaded Weight	23,900 kg
Cylinder Arrangement	12, V-Shape 60 deg.		Cross-country — 95 km	Payload	2,000 kg
Bore	105 mm	Turning Radius	5.85 m	Crew	4 Man
Piston Displacement	11,867 cc	Suspension	Torsion Bar, Transverse, 12 Bars	Fuel Consumption	200 l per 100 km of Road
Compression Ratio	6.5 : 1	Wheel Base	2,510 mm	Fuel Capacity	310 l
RPM Normal/Maximum	2,600/3,000	Length of Track in Contact with Ground/Number of		Armor	
Maximum Power	265/300 h.p.	Links	2860 mm — 93 links	Forward Hull	50 + 30 mm
Valves	Hanging	Track Width	400 mm	Sides and Rear	30 mm
Crankshaft Bearings	7 Roller	Ground Clearance	390 mm	Performance	
Carburettors	2 Solex Type 40 JFF II	Overall Length	6,770 mm	Gradient	30 deg
Batteries	2 x 12 Volt 105 Amp.	Overall Width	2,950/3,330 mm with Ostkette	Vertical Climb	600 mm
Number of Speeds	Forward 6, Reverse 1		3,410 mm with Schürzen	Wading Depth	800 mm
Maximum Speed	40 kph	Overall Height	2,160 mm	Gap Crossing Ability	2,300 mm
		Axis Height of Gun	1,570 mm	Arm. and Ammo	75mm Stuk 40 L/48 (54 rounds)
		Ground Pressure	1.04 kg/cm2		1 — MG 34 (600 rounds)
		Vehicle Weight	15,800 kg		

Above: There were also stylish *Schürzen*.

Right: Sturmgeschütz or Sturmhaubitze?

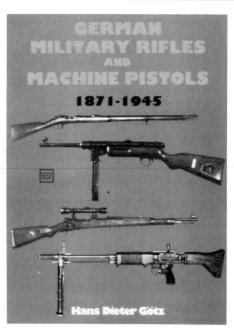
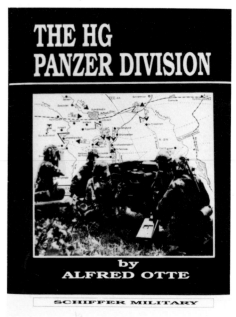

· Schiffer Military History ·

Specializing in the German Military of World War II

Also Available:

• The 1st SS Panzer Division - *Leibstandarte* • The 12th SS Panzer Division - *HJ* • The Panzerkorps *Grossdeutschland* •
• The Heavy Flak Guns 1933-1945 • German Motorcycles in World War II • Hetzer • V2 • Me 163 "Komet" •
• Me 262 • German Aircraft Carrier *Graf Zeppelin* • The Waffen-SS - A Pictorial History • Maus • Arado Ar 234 •
• The Tiger Family • The Panther Family • German Airships • Do 335 •
• German Uniforms of the 20th Century - Vol.1 The Panzer Uniforms, Vol. 2 The Uniforms of the Infantry •

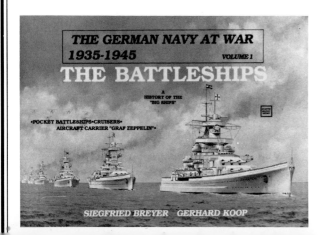